Innovations and Practices in Fisheries and Aquaculture

NIPA® GENX ELECTRONIC RESOURCES & SOLUTIONS P. LTD.
New Delhi-110 034

Innovations and Practices in Fisheries and Aquaculture

Pathways for Achieving Sustainable Development Goals

B.K. Das
S. Samanta
Anjana Ekka
Sajina A.M.
Sangeetha M. Nair
Mishal P.
A.K. Das

NIPA® GENX ELECTRONIC RESOURCES & SOLUTIONS P. LTD.
New Delhi-110 034

NIPA® GENX ELECTRONIC RESOURCES & SOLUTIONS P. LTD.

101,103, Vikas Surya Plaza, CU Block
L.S.C. Market, Pitam Pura, New Delhi-110 034
Ph : +91-11-43860225, Mob.: +91 9717133558, 9540816132
E-mail: newindiapublishingagency@gmail.com
Website: www.nipaersources.com

Print ISBN: 978-93-58877-10-6

ebook ISBN: 978-93-58871-76-0

Composed and Designed by NIPA®.

Preface

Fisheries and aquaculture play a crucial role in food security, nutrition, and livelihoods, particularly in a country like India, with vast marine and inland water resources. As the second-largest producer of fish globally, India's fisheries sector contributes significantly to employment, income generation, and economic growth. However, the sector also faces pressing challenges such as overfishing, habitat degradation, climate change, and socio-economic disparities among fishing communities. To address these challenges while harnessing the full potential of the sector, innovative and sustainable practices must be adopted in alignment with the Sustainable Development Goals (SDGs).

This book explores pathways to achieving sustainability in fisheries and aquaculture by integrating scientific advancements, traditional ecological knowledge, and policy interventions. With a special focus on India, it delves into topics such as ecosystem-based fisheries management, technological innovations in aquaculture, community-led conservation initiatives, and climate-resilient strategies. It also highlights the role of women and marginalized communities in the fisheries value chain, advocating for inclusive and equitable growth.

Through case studies, empirical analyses, and policy discussions, this work aims to provide a comprehensive understanding of how sustainable fisheries and aquaculture can contribute to multiple SDGs, including zero hunger (SDG 2), life below water (SDG 14), decent work and economic growth (SDG 8), and climate action (SDG 13). By fostering interdisciplinary collaboration among researchers, policymakers, practitioners, and local communities, we can chart a course toward a more resilient and equitable blue economy.

It is hoped that this volume will serve as a valuable resource for academics, students, policymakers, and stakeholders committed to the sustainable development of India's fisheries and aquaculture sector.

Editors

Contents

1

Fostering Inland Open Water Fisheries of India Towards Attaining Sustainable Development Goals (SDGs)

Basanta Kumar Das

ICAR-Central Inland Fisheries Research Institute, Barrackpore, Kolkata-700120 West Bengal, India

Email: director.cifri@gmail.com

Introduction

India possesses a diverse array of aquatic ecosystems, including rivers, lakes, reservoirs, wetlands, lagoons and estuaries, which are vital for sustaining a thriving inland open water fisheries sector. This sector is instrumental in providing livelihoods to millions of people, reinforcing food security, and fostering socio-economic development. Aligned with the Government of India's Viksit Bharat@2047 vision, aimed at transforming the nation into a developed state by its centennial year of independence in 2047, efforts are focused on enhancing fish production. The Prime Minister's Matsya Sampada Yojana (PMMSY) has been pivotal in leveraging technology and research to propel the fisheries sector forward in the country. Presently, India ranks as the world's third-largest fish-producing nation, contributing 15.8 million metric tonnes (FAO, 2024a), with the inland sector accounting for 75% of the nation's total fish production and supporting over 9.45 million livelihoods. India leads globally in inland capture fish production (1.9 million metric tonnes) (FAO, 2024b), underscoring its substantial potential in advancing sustainable management of inland water resources towards meeting food and livelihood security in the context of emerging challenges including climate change. ICAR-CIFRI technologies and management measures are playing a key role in sustainable fish production enhancement and management of inland open water sector of the country in the last 7 decades.

The abundant inland open water resources offer immense potential for fish production and livelihood support for rural communities. Despite this potential, current fish yield from reservoirs and wetlands remains low, presenting an opportunity for advancement through improved management and technological backstopping. Sustainable growth in inland open water fisheries can be achieved through strategic interventions such as culture-based fisheries in floodplain wetlands and small reservoirs, stock enhancement in medium and large reservoirs as well as conservation efforts in rivers and utilization of underutilized resources like canals, coal pits, minor irrigation tanks, check dams and upland lakes. Enclosure culture systems (cage & pen) offer promise in increasing fish output while ensuring environmental conservation. India's inland open water fisheries are characterized by a diverse array of freshwater fish species. The major river systems such as the Ganges, Brahmaputra, Godavari, Krishna, and Cauvery, along with their tributaries, support diverse fish populations and nutritional security. The distribution of fish species varies across different regions of the country, influenced by factors such as climate, water quality, and habitat availability. Effective management of riverine fisheries is essential for conserving biodiversity, ensuring food security, supporting livelihoods, preserving cultural heritage, maintaining ecosystem services, promoting tourism, and building resilience to climate change. Collaboration among stakeholders, including government, local communities, and the private sector, is crucial for the sustainable management of India's inland fisheries resources.

Inland open water fisheries play a significant role in contributing to several Sustainable Development Goals (SDGs). Firstly, they contribute to SDG 1 (No Poverty) by providing livelihood opportunities for millions of people, especially in developing countries. Secondly, they support SDG 2 (Zero Hunger) by providing a vital source of protein and essential nutrients to communities worldwide, particularly in areas where alternative food sources are scarce. Additionally, sustainable management of inland open water fisheries aligns with SDG 14 (Life Below Water) by promoting responsible practices that conserve aquatic ecosystems and biodiversity. Moreover, by generating income and improving food security, inland open water fisheries contribute indirectly to other SDGs such as SDG 8 (Decent Work and Economic Growth) and SDG 10 (Reduced Inequalities). Through leveraging innovation, technology, and responsible governance, the sector can meet several SDGs by unlocking the potential and economic value of vast inland aquatic resources. Therefore, recognizing the importance of sustainable management practices in inland open water fisheries is crucial for achieving multiple SDGs and promoting holistic development.Despite facing challenges such as habitat degradation, overfishing, pollution, climate change and invasive species, India's inland open water fisheries remain vital for supporting fisheries and livelihood. Conservation efforts are underway to address these challenges, including habitat restoration, regulation of fishing practices, promotion of sustainable fisheries and enclosure culture, community empowerment, strengthening value chain, robust governance and funding support aimed at fostering sustainability and inclusivity. The sector can play a crucial role in providing employment opportunities and nutritional security, thus emphasizing the importance of their conservation and sustainable management for attaining SDGs.

Inland openwater fisheries resources of India

The inland fisheries resources of India are diverse and dispersed across different agro-climatic region of the country. The primary inland open water resources are rivers, reservoirs, wetlands, lakes, lagoons, estuaries and associated water bodies. There is a tremendous potential for fish production with these fisheries resources. However, utilizing their fisheries potential is quite challenging due to multitude of factors that requires adoption of scientific management strategies coupled with fund flow, robust governance and policy support.

Table 1.1: Inland open water resources of India

Resources	Area (in million ha.)
Rivers	29,000 km (major), 164,118 km total
Mangroves	0.356
Estuaries	0.458
Estuarine wetlands	0.04
Backwaters/ Lagoons	0.246
Reservoirs	3.460
Floodplain wetlands	0.564
Upland lakes	0.091

Rivers

The Ganga River System, the Brahmaputra River System, the Indus River System, the East Coast River System, and the West Coast River System are among five major river systems in the country. There are 14 main rivers, 44 medium rivers, countless minor rivers, and rivulets make up the 29,000 km long network of major rivers.The major river basins alone cover 83% of the total drainage area and contribute about 85% of the overall surface runoff, fulfilling the needs of over 80% of the population. Indian rivers harbors one of the richest fish genetic resources globally, with the Gangetic system alone supporting 265 species, the Brahmaputra system hosting 126 species, and peninsular rivers nurturing 76 fish species. Regular systematic recording and reporting of data on fish catch from rivers are lacking. The data collected by ICAR-CIFRI from selected stretches of rivers including the Ganga, Brahmaputra, Narmada, Tapti, Godavari, and Krishna reveal a wide variation in fish yield, ranging from as low as 5 kg/km to as high as 3,300 kg/km. The rivers sustain a mix of artisanal, subsistence, and traditional fisheries, despite facing challenges from a highly dispersed and unorganized marketing system. The current distribution of water usage from river is concerning for sustainable fisheries, as over 80% is allocated for irrigation, followed by domestic and other sectors. There's an anticipated rise in water demands across these sectors, which will further strain river flows. Conservation and restoration efforts are crucial for the overall sustainable development, as issues such as altered and diminishing flow, habitat degradation, pollution, declining fish catch, loss of fish diversity, influx of exotics and destructive fishing practices threaten these invaluable resources.

Estuaries

Estuarine systems, spanning 4,58,185 ha, are vital fisheries resources, serving as crucial breeding grounds for commercially important fish and shellfish. Their conservation and management are of paramount importance due to their significant contribution beyond subsistence levels, with an average yield of approximately 400 kg/ha, sustaining the fishery sector. However, modifications to river courses have adversely affected fish populations. There's a dual challenge: some estuaries may become hyper-saline due to reduced freshwater inflow, while others risk losing established management protocols, endangering fish stocks and biodiversity. While research has been conducted, continuous monitoring is mainly established in the Hooghly-Matlah estuarine system by ICAR-CIFRI. Balancing exploitation and conservation are essential for the sustainable utilization and preservation of these critical ecosystems.

Lagoons and backwaters

Lagoons and backwaters linked with estuaries represent a significant inland capture fishery resource. Notably, Chilika (1,16,500 ha) and Pulicat Lake (79,000 ha) along the east coast, as well as a series of lagoons (46,129 ha) in Kerala, including the Vembanad lagoon on the west coast, stand out as major brackishwater resources in India. However, diminished river flows, siltation, and human-induced pressures in the surrounding catchment areas have imposed substantial stress on the environment and fishery within these water bodies. Addressing these challenges is imperative to ensure the sustainability of these valuable inland fisheries resources. The institute is assessing the time scale changes in these resources and continuously providing strategies for fisheries management.

Reservoirs

India possesses approximately 3.46 million hectares of reservoirs, predominantly located in the peninsular and central Indian states with small reservoirs occupying 36% (12,72,821 ha), medium reservoirs covering 16% (5,75,290 ha), and large reservoirs spanning 47% (16,12,190 ha) of this area. These reservoirs play a significant role in contributing to inland open water fish production. However, despite their vast potential, reservoir fisheries in India remain largely underutilized though there has been substantial increment in fish yield over the years following ICAR-CIFRI guidelines. One of the primary challenges hindering the effective harnessing of reservoir fisheries potential is the lack of access to high-quality fish seed. Furthermore, the stocking of fish in Indian reservoirs has often been conducted without adequate consideration of scientific principles. Addressing these challenges requires concerted efforts to improve access to high-quality fish seed, enhance financial support for reservoir fisheries management, and adopt science-based approaches to stocking and management practices. By addressing these issues, India can unlock the potential of its reservoir resources and promote open water fish production.

Floodplain wetlands

Floodplain wetlands (0.564 million ha) are one of the most productive inland open water fisheries resources distributed primarily in Eastern and North Eastern India. These wetlands are characterized by their dynamic nature, being periodically inundated by seasonal floods and supporting rich biodiversity (96 fish species). In India, floodplain wetlands are found along the floodplains of major rivers such as the Ganges, Brahmaputra, and their tributaries. These wetlands play essential ecological roles, including flood regulation, groundwater recharge, and sediment deposition, while also providing important breeding grounds for fish and other aquatic species. Additionally, floodplain wetlands contribute to the livelihoods of local communities through fishing, agriculture, and ecosystem services such as water purification and nutrient cycling. Despite their ecological significance, floodplain wetlands in India face multiple challenges. Conservation and sustainable management efforts are essential to preserve these valuable ecosystems and ensure their continued contribution to biodiversity conservation, water resource management, and community well-being.

Table 1.2: Floodplain wetlands resources of India

State	River basin	Area (ha)
West Bengal	Ganga (Bhagirathi & Hooghly), Jalangi, Ichamati, Padma, Matlah and Mahananda	42,500
Bihar	Gandak and Koshi	2,40,000
Uttar Pradesh	Ganga and Yamuna	1,50,000
Assam	Brahmaputra and Barak	1,00,000
Manipur	Iral, Imphal and Thoubal	30,171

Status and potential of fish production from reservoirs and wetlands

In Indian reservoirs, Culture-Based Fisheries (CBF) has been successfully implemented by stocking seeds of Indian Major Carps. The eco-friendly CBF strategies developed by ICAR-CIFRI have been widely embraced in various reservoirs, leading to significant advancements in fish production. During the 1990s, the average annual fish yield from Indian reservoirs was notably low, with figures of 49.9, 12.3, and 11.4 kg/ha/year for small, medium, and large reservoirs, respectively, compared to their potential yields of 100, 75, and 50 kg/ha/year. Recognizing the substantial untapped potential, the Government of India launched fingerling stocking programs through different schemes with technological backstopping from ICAR-CIFRI. As a result of these initiatives, there was a marked increase in fish production, with the average fish yield rising from 20 to 110 kg/ha/year. Through the successful adoption of fisheries management practices, including CBF, the fish yield has seen a substantial increase to 190, 98, and 34 kg/ha/year against a recasted potential of 500, 200, and 100 kg/ha/year from small, medium, and large reservoirs, respectively. Despite this progress, current fish production stands at 3.86 lakh tonnes (112 kg/ha), a fraction of its potential, which is estimated at 1million metric tonnes, highlighting the substantial gap between current production levels and the reservoirs'capacity. Under PMMSY, projections aim to elevate the average fish yield to 343 kg/ha/year for small reservoirs, 147 kg/ha/year for medium reservoirs, and 50 kg/ha/year for large reservoirs.

Table 1.3: Potential & projected fish yield in reservoir through CBF & stocking enhancement

Reservoir category	Present fish yield (kg/ha/yr)	Potential yield (kg/ha/yr)	Yield gap (kg/ha/yr)	Targeted yield (kg/ha/yr)
Small	190	500	310	343
Medium	98	200	102	147
Large	34	100	66	50

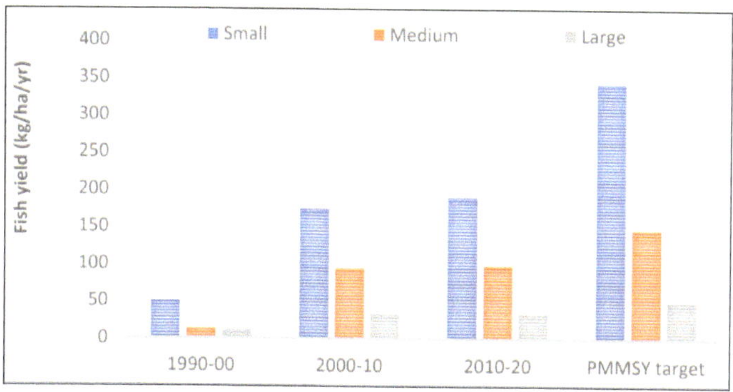

Fig. 1.1: Increment in fish yield of reservoirs over the years

Floodplain wetlands in India, with a production potential of 1500 to 2000 kg/ha/year, have witnessed a remarkable increase in average fish yield due to the adoption of CBF practices. This transformation has seen the average yield surge from 50 kg/ha/year to over 400 kg/ha/year. Notably, West Bengal stands out with an impressive average yield surpassing 700 kg/ha/year, largely attributed to robust governance facilitated by active fishermen cooperatives. Assam, Uttar Pradesh, and Bihar have also experienced significant increases in fish yield, reaching impressive levels of 450, 350, and 300 kg/ha/year, respectively. These notable improvements highlight the positive impact of effective management strategies, including CBF, in enhancing the productivity of floodplain wetlands. However, the present production is around 2.0 lakh tonnes against a potential of 5 lakh tonnes leaving scope for further enhancement of fish production.

Table 1.4: Fish yield and potential of floodplain wetlands

State	Average yield (kg/ha/yr)	Potential yield (kg/ha/yr)
West Bengal	700	1500
Bihar	300	800
Uttar Pradesh	350	800
Assam	450	1500
Manipur	200	500

Recent interventions in Bihar wetlands

ICAR-CIFRI's technical interventions have significantly influenced the fisheries of the five adopted wetlands in East Champaran, Bihar, yielding remarkable outcomes. Through the comprehensive implementation of the CBF protocol, establishment of infrastructure, and providing training and exposure visits to fishers, fish production has doubled besides increase in fishing days. In Kararia, fish yield surged from 190 kg/ha/yr to 592 kg/ha/yr, accompanied by improved catch per unit effort, with fishing man days increasing from 44 to 151 days annually. Majharia witnessed an increase in fish yield from 60 to 87 kg/ha/yr, with fishing man days rising from 33 to 66.25 per year. Rulhi Maun experienced a notable enhancement in fish production, with yield climbing from 130 kg/ha/yr to 300 kg/ha/yr. Significantly, in the Sirsa wetland, the interventions led to a remarkable doubling of fish production from 230 to 432 kg/ha/yr, along with an increase in fishing days from 32 to 62, and a notable rise in fishers' income over just two years.

Small scale inland open water fisheries

In India, the majority of inland open water fisheries (>90%) are characterized as small-scale fisheries (SSF). These SSF play a crucial role contributing significantly to the hidden harvest, accounting for up to 30% of total fish production in various resources. These fisheries are vital for addressing food security challenges, improving household nutrition, and sustaining livelihoods, particularly in rural and coastal communities. Given their importance, there is a special focus on SSF by the institute in project mode as a key engine for promoting food security, enhancing nutrition, and supporting sustainable livelihoods. By prioritizing the development and support of SSF, India can not only boost its fisheries sector but also contribute to achieving Sustainable Development Goals (SDGs), particularly those related to poverty alleviation, food security, and economic growth. Therefore, investing in the growth and resilience of SSF is essential for ensuring the well-being of communities reliant on inland open water fisheries across the country. ICAR-CIFRI in collaboration with WorldFish is conducting research and developmental activities in different wetlands of West Bengal and Assam yielding significant achievements.

Inland open water fisheries and SDGs

The inland fisheries sector is important for achieving the SDGs around the world. This stems from the fact that this sector provides food and nutritional security to billions of people and caters to livelihood functions of millions of people in the world. ICAR-CIFRI emphasizes on harnessing the potential of inland open water fisheries and interventions to address the SDGs through various flagship programmes of the institute. With this aim in view, institute has worked on SDG 1 (No Poverty: Reduce Poverty by Involving PFCS, SHGs, NGOs etc.); SDG 2 (Zero Hunger: Ensure Household Nutritional Security through SIFs); SDG 5 (Gender Equality: Women Empowerment through Ornamental Fish Culture); SDG 8 (Decent Work and Economic Growth: Increase Farmers' Income through Production Enhancement Strategy); SDG 12 (Responsible Consumption and Production: Sustainable Fish Production); SDG 13 (Climate

Action: Culture and Popularize Climate Resilient Fish Species); and SDG 14 (Life Below Water: Conserve Fish Diversity with Special Reference to SIFs). These interventions are not only helping in achieving individual targets but also resulting in mutually beneficial synergies across the SDGs.

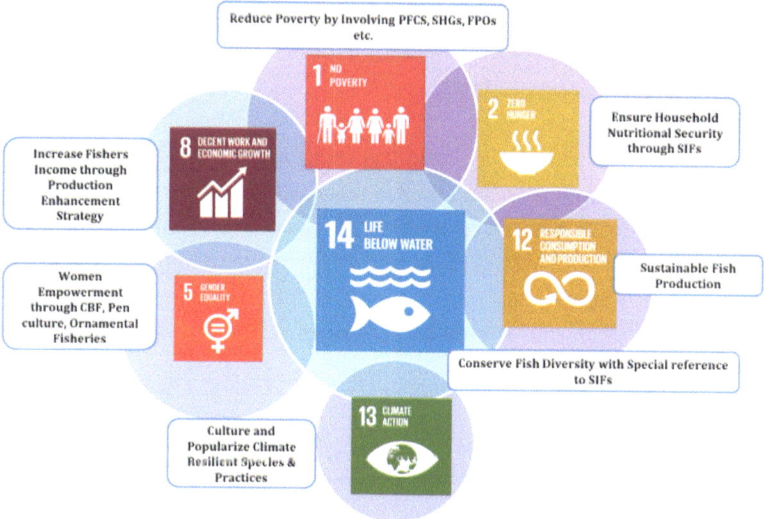

Fig. 1.2: SDGs addressed by ICAR-CIFRI through research and developmental activities

Importance of SIFs

Among the freshwater fish in India, 450 are classified as SIF species. In river Ganga, 63 out of 190 fish species are SIFs, with Cypriniformes being the most prevalent order. In the Ganga basin, common SIF groups include minnows, barbs, schilbeids, clupeids, and bagrids. These species contribute significantly to the river's indigenous fish population, with *Cabdio morar, Salmostoma bacaila*, *Puntius sophore, Gudusia chapra, Securicula gora, Parambassis ranga, Osteobrama cotio* and others being particularly abundant. Similarly, around 60 SIF species are found in the River Godavari, with barbs and minor carps being dominant. Wetlands play a vital role in supporting SIF populations (around 40 species), providing habitats for breeding and feeding. SIFs are essential for local communities, contributing to both diets and livelihoods through subsistence fishing and aquaculture. SIFs are often overlooked in catch statistics as they are mainly consumed locally, contributing to hidden harvests and household nutrition. Highlighting their importance, specific SIF species with high market demand have been identified by various researchers to mitigate micronutrient deficiencies. The database NutriFishIn developed by ICAR-CIFRI contains the nutritional information on food fishes of India and it helps to identify species rich in various micronutrients. Utilizing SIFs could boost fishing communities' income, improving access to education, health, and nutrition catering to several SDGs.

Issues in open water fisheries

Challenges in the riverine sector include impaired flows, depletion of fish stocks, overexploitation, biodiversity decline, invasion of exotic species, habitat degradation, pollution (heavy metals, pesticide residues), encroachment, siltation, lack of reliable fish catch data, poor socio-economic status of fishers, and weak implementation and monitoring of fisheries regulations. Additionally, many prized fish species are trans-boundary, lacking a common management strategy. Estuaries face hyper salinity from reduced freshwater influx or loss of saline character due to excessive freshwater discharge, impacting species diversity, fish catch, and seasonal availability. Climate change exacerbates these issues, posing a significant threat to estuarine fisheries. Reservoir and wetland fisheries also face multitude of challenges which are described below in figures.

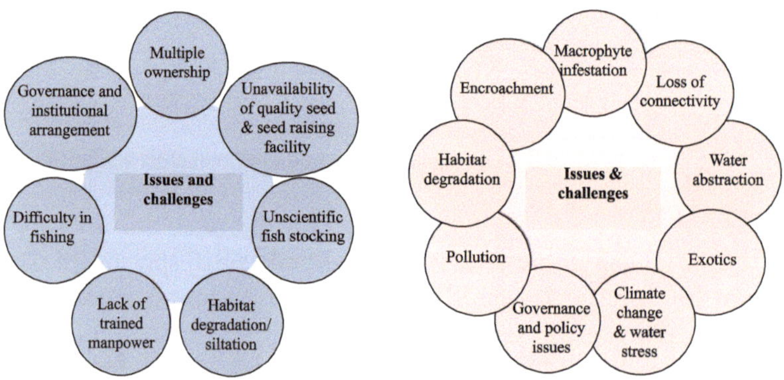

Fig. 1.3: Challenges and issues of reservoir (A) and floodplain wetland fisheries (B)

Management strategies for inland open water fisheries

Reservoir fisheries management

Culture-based fisheries (CBF) in small reservoirs

The main goal of CBF is stocking and recapture, where fish survival and growth are both influenced by stocking density. Success in CBF depends on species choice, fishing effort, size at capture (species dependent), stocking size and timing, stocking density, and fishing gear. Due to their smaller size, shallower depth, and easy recapture of stocked fish, irrigation tanks and canals are also appropriate for CBF.The average fish yield of 2300 reservoirs increased significantly from 20 to 110 kg/ha/year through the adoption of CBF technology developed by ICAR-CIFRI.

Stock enhancement in medium and large reservoirs

The main goal of stock enhancement is to improve/supplement the self-recruiting populations. When production is lower than the potential, stock enhancement is the main strategy utilized to sustain or improve fish stocks in open water bodies like large reservoirs. Knowledge of population structure, including recruitment, growth, and mortality of the stock, are essential for the effectiveness of stock enhancement. One of the essential conditions for effective reservoir fishery management is the augmentation of the stock to prevent undesirable fish from utilizing the available food niches and flourishing at the expense of economically significant fast-growing species that colonize all the diverse niches of the biotope. The environment should be conducive for the growth and reproduction of the species, and it should have fair chances of establishing in the water body. The research by the institute have revealed success of stock enhancement regime in various large reservoirs of the country.

Impact of stocking in reservoirs

The implementation of CBF across 58 small reservoirs in Odisha has resulted in a significant increase in fish production, rising from 204 to 323 kg/ha/year. This has led to a notable enhancement in per capita fish output, rising from 398 to 702 kg per fisher annually. In selected reservoirs located in Chhattisgarh, Madhya Pradesh, Jharkhand, Karnataka, and Tamil Nadu, interventions by ICAR-CIFRI have led to a fish production level of 125 kg/ha/year. Similarly, the fish productivity of the large Gandhisagar reservoir in Madhya Pradesh has reached 100 kg/ha/year due to stocking. The Karapuzha reservoir in Kerala has seen a remarkable surge in fish yield, progressing from 18 to 150 kg/ha/year. Notably, strategic stocking has resulted in a substantial increase in fish yield in large reservoirs of Madhya Pradesh, from 12.3 to 52.4 kg/ha/year. Moreover, the adoption of CBF has led to a noteworthy rise in fish yield across 24 reservoirs in Chhattisgarh, escalating from 80.43 kg/ha/year to an impressive 374.46 kg/ha/year.

Wetland fisheries management

CBF is the most widely accepted technology for enhancement of fish production in wetlands. The adoption of CBF in wetlands of the eastern and north-eastern states of the country has significantly increased the fish production. Due to the adoption of CBF regime developed by ICAR-CIFRI, the fish yield from floodplain wetlands have increased from 50 kg/ha/year to 400 kg/ha/year and presently yielding around 2 lakh tonnes of fish. Significant increase in the fish yield from 234.51 kg/ha/year to 704.60 kg/ha/year was observed in the wetlands of North East India due to adoption of CBF. Production level as high as 2000 kg/ha/year was also achieved in selected wetland of West Bengal with the technical support of ICAR-CIFRI.

Pen culture

The integration of pen culture developed by the institute into wetland fisheries management has proven to be a transformative approach, yielding numerous benefits. Notably, it significantly reduces the cost of producing quality fish seed, with impressive results such as generating 500 kg of carp fingerlings exceeding 100 mm in size from just 0.1 ha of pen. Moreover, this technique has demonstrated remarkable potential for table size fish production, achieving yields ranging from 3000 to 4000 kg/ha/8months. The success of this ICAR-CIFRI technology has been widespread, with its demonstration across the country, particularly gaining traction in wetlands located in Assam and West Bengal. Additionally, the utilization of derelict, weed-choked areas for pen culture presents an innovative solution, effectively repurposing challenging landscapes where traditional fishing methods are difficult. Overall, the integration of pen culture into wetland fisheries management stands as a promising strategy, offering cost-effective fish seed production, increased table fish yields, and the revitalization of underutilized aquatic habitats.

Ecosystem-based approach

Adopting an ecosystem-based approach is crucial for managing open water fisheries and expanding the dissemination of CBF to broader geographical areas. This approach adopted and customized by ICAR-CIFRI involves utilizing ecosystem modeling to manage fisheries comprehensively, considering the interconnectedness between different species and their environment within a changing scenario in inland open waters of India. Employing modeling techniques for stocking management in CBF/stock enhancement can improve the efficient utilization of available resources, leading to optimal production outcomes. It's imperative to avoid unscientific stocking practices, which can result in significant production losses when implementing CBF. Leveraging advanced modeling techniques can further enhance the ability to simulate the dynamics of fish populations, predict the potential impacts of management decisions, and optimize harvesting strategies for sustainable yields. Conducting in-depth studies on the biological characteristics of fish species is essential to inform effective management practices and conservation efforts. By integrating ecosystem modeling and advanced biological characterization studies, fisheries management can be better equipped to navigate complex ecological dynamics and promote sustainable practices.

Ecological modeling

Ecological modeling is a scientific approach used to simulate and understand the complex interactions within ecosystems. These models can range from simple conceptual frameworks to sophisticated simulations based on detailed data and equations. Ecological modeling allows researchers to explore hypothetical scenarios, predict the effects of environmental changes, and assess the impacts of human activities on ecosystems. By integrating ecological theory with empirical data and computational techniques, ecological modeling provides valuable insights into ecosystem functioning, biodiversity conservation, and sustainable management practices. Furthermore, it plays a crucial role in the decision-making processes for environmental policy, resource management, and conservation efforts. ICAR-CIFRI has developed mass balance models, such as Ecopath with Ecosim, for fisheries management in peninsular reservoirs and the Chilika lagoon. These models were also utilized to propose optimal stocking density and size at stocking for a reservoir in Kerala. Recently, the Institute has developed ecosystem-based model for the floodplain wetland fisheries management.

Tapping potential of unutilized water bodies

Harnessing the potential of underutilized resources such as canals, coal pits, upland lakes, check dams, and minor irrigation tanks represents a significant opportunity to enhance fish yields. These water bodies often remain overlooked or abandoned, yet they hold immense potential for supporting aquatic ecosystems and boosting fish production. By repurposing these underutilized resources for enhancement activities, such as stocking with fish fingerlings, proper management, and habitat restoration, we can transform them into valuable assets for sustainable fisheries development.

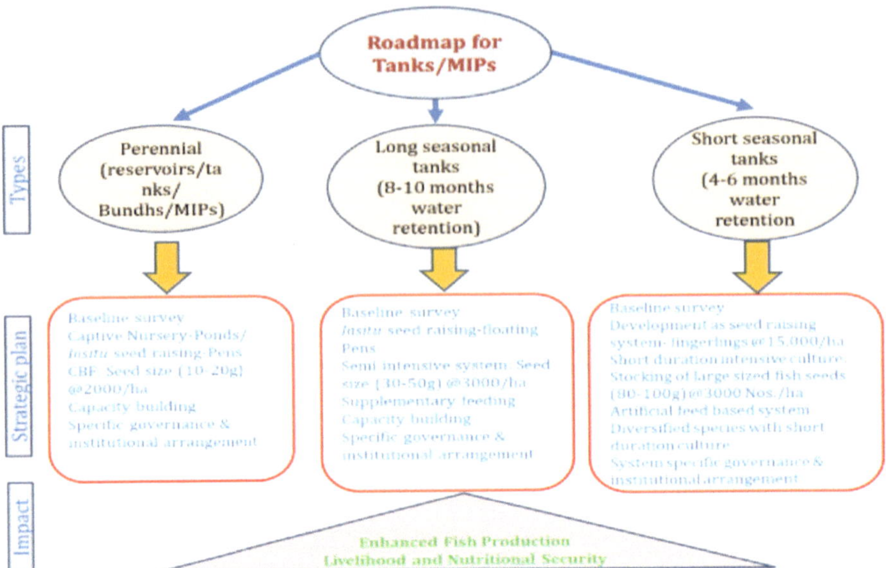

Fig. 1.4. Roadmap for tapping the fisheries potential of Tanks/Micro Irrigation Projects

Riverine fisheries management

River ranching

The major objective of river ranching programs is to replenish and revive vulnerable and declining native fish populations by introducing fish seed into the surrounding ecosystem. This fish seed can be sourced either through the collection of wild seeds from the same river system or by artificially raising fish. It is crucial to take precautions during the ranching program to prevent genetic contamination. To maximize the survival rate of ranched fish, specific locations along a stretch of river should be identified for ranching, and fishing activities in the surrounding areas should be restricted or limited. In order to successfully implement river ranching, a consistent supply of high-quality fingerlings of desired fish species is essential. Recognizing this need, ICAR-CIFRI has established on-field broodstock development and fingerling-raising facilities. Since 2016, CIFRI has been spearheading a scientific river ranching program in the Ganga River under the National Mission for Clean Ganga (NMCG) initiative of the Government of India.

In a significant effort toward fisheries conservation and restoration, ICAR-CIFRI has organized approximately 100 river ranching programs across India. These initiatives have led to the release of 143 lakh (generated through induced breeding of Gangetic brooders) IMC fingerlings, Mahseer, and other native fishes into the river Ganga, spanning five different states. Recognizing the importance of such endeavors, CIFRI has launched a mission mode National Ranching Programme Campaign aimed at bolstering fish conservation efforts nationwide. West Bengal has seen the highest proportion of ranching activities at 51.8%, followed by Uttar Pradesh (27.8%), Bihar (12.8%), Jharkhand (7.4%), and Uttarakhand (0.20%). Over the years, ranching activities have witnessed a significant increase, with the number of fingerlings released rising from 2.0 lakhs in 2020 to an impressive 30 lakhs in 2023. These endeavors underscore the commitment to safeguarding fish populations and fostering sustainable fisheries management practices across the region.

Habitat protection and ban season

Habitat protection within river systems stands as a crucial management practice aimed at conserving fish diversity and fostering overall fisheries development. By imposing restrictions on fishing and other anthropogenic activities in protected areas, a natural environment can be provided for fish breeding and the growth of juveniles. Designating deep pools within rivers as sanctuaries has proven effective, as these areas harbor a variety of species that serve as a seed multi-store for the entire river stretch. Additionally, implementing a ban season represents another management measure geared towards conserving fish diversity and supporting fisheries development. During the ban period, fishing activities, especially during the breeding season, has to be prohibited to prevent disruption of fish recruitment activities. It's essential for the ban period to align with the peak breeding periods of various fish species to maximize its effectiveness in safeguarding fish populations and promoting sustainable fisheries management practices.

Hilsa conservation

Extensive studies have been conducted by the institute on the biology, breeding, population dynamics, and migration patterns of the Hilsa, a prized fish species in India. The study conducted on the Hilsa fishery management in the river Hooghly revealed a concerning 20% overexploitation of the Hilsa stock. To ensure sustainable management of the Hilsa fishery, comprehensive guidelines have been developed and provided to the West Bengal government. These include implementing a minimum 20% reduction in fishing effort, imposing restrictions on the use of gill nets with a mesh size less than 90 mm, and enforcing a ban on fishing of brood fish during the breeding season, which spans from June to August. These measures aim to mitigate the pressures on the Hilsa population and promote sustainable fishing practices, thereby safeguarding the long-term viability of the fishery.The institute is instrumental in the breeding of Hilsa and rearing for its conservation during the last several decades with significant strides in research. As part of hilsa conservation efforts, a Hilsa ranching station has been established at Farakka. A significant milestone was achieved with the ranching Hilsa brood fish upstream of the barrage. Moreover, over 8.5 lakh fertilized eggs and spawn were released at various sites upstream of the Farakka Barrage, bolstering efforts to replenish Hilsa populations in the region. Additionally, an extensive awareness campaign has been conducted, sensitizing 8,326 fishers across Uttarakhand, Uttar Pradesh, Bihar, Jharkhand, and West Bengal emphasizing the importance of conservation measures and responsible fishing practices. These initiatives reflect a concerted effort to safeguard the Hilsa population and promote the sustainable management of this valuable fishery resource.

Pollution monitoring and abatement

The main sources of pollution in rivers are industrial and domestic wastes. Installation of effluent treatment plants (ETPs) and sewage treatment plants (STPs) in major cities and towns along rivers is critical for pollution control, maintaining ecosystem health and safety of fish for consumption. Emerging contaminants in rivers represent a growing concern worldwide, reflecting the evolving landscape of pollutants entering freshwater ecosystems. These contaminants encompass a diverse range of substances, including pharmaceuticals, personal care products, industrial chemicals, pesticides, and microplastics, among others. Unlike traditional pollutants, emerging contaminants may not be routinely monitored or regulated, posing challenges for water quality management. Hence,ICAR-CIFRI is conducting research on emerging contaminants including pesticides, heavy metals, EDCs and micro-plastics in inland open waters and the institute has benchmarked the status of pollutants in major rivers of India.

e-flow assessment

The effective utilization of environmental flows as a decision support tool in India was limited, mirroring a broader trend seen in many developing nations. However, there has been a recent surge in awareness surrounding environmental flows, prompting policymakers to act. Many available models do not adequately address the complexities of Indian river systems. Altering a river's natural flow regime can disrupt the entire river ecosystem, resulting in the collapse of riverine fisheries and fish diversity. As a result, Govt. of India is increasingly prioritizing efforts to address these issues and incorporate environmental flow considerations into river management policies and practices. ICAR-CIFRI is playing a pivotal role by conducting research and recommending e-flow requirements in the context of emerging river valley projects for sustainable aqua life of Indian rivers.

Enclosure culture

Cage and pen have greater scope for the production of table size fish thereby increasing the per unit water productivity of reservoirs and wetlands. There is a huge demand of fish seed for CBF and hence *in situ* raising of fish seed either in cages or pens could be a viable and cost-effective option to support the stocking programme in wetlands and reservoirs. ICAR-CIFRI is the pioneer for the development of technology of inland open waters of India. Technologies developed by the institute on cage culture have been applied successfully in many reservoirs in different parts of the country. Due to vast reservoir resources, more than 2 mmt can be easily produced by utilizing a small fraction (0.1%) of the total area of medium and large reservoir in the country. In view of above, the cage culture may be a viable technology for fisheries enhancement in inland open waters. Due to its successful result, more than 20000 cages have been installed in different parts of India and additional 20000 cages have been proposed in PMMSY.

Table 1.5: Potential of cage culture in medium and large reservoirs

Total area (medium and large) in ha	0.1% area (ha)	Total number of cages	Potential production (tonnes)
2157318	2157	862927	2588782

Species and system diversification

The selection of species for cage culture should be tailored to meet market demands, considering factors such as market value, the resilience of the species, and its ability to thrive with external sources of food in confined conditions.

While economically viable cage culture of *Pangasianodon hypophthalmus* is practiced in the country's reservoirs, there is a need to encourage cage culture of more species from the indigenous pool. Considering the consistent demand for economically and nutritionally valuable species, and regional preferences for specific species, the following indigenous species can be considered for cage culture based on the research finding of ICAR-CIFRI: *Labeo bata, L. rohita* (Jayanti rohu), *L. gonius*, *Osteobrama belangeri* (Pengba), *Systomus sarana, Ompok bimaculatus* (Pabda), *Etroplus suratensis, Macrobrachium* sp., etc. It is essential to note that, apart from *Pangasianodon hypophthalmus* and GIFT tilapia, the cage culture of any other exotic species, including those illegally introduced, is strictly prohibited in inland open waters by the guideline of NFDB.

The cage culture technology for *Pangasianodon hypophthalmus* in reservoirs has been standardized, achieving an average production of 2-3 tonnes/ cage measuring 5m x 5m x 4m. ICAR-CIFRI has commercialized GI cages suitable for both fish seed raising and table fish production. Recognizing the immediate necessity for diversification to enhance adoptability and profitability, the feasibility of several commercially important indigenous fish species, including *Labeo rohita* (Jayanti), *Ompok bimaculatus, Labeo bata, Labeo gonius, Barbonymus gonionotus*, and *Systomus sarana* in different agro-climatic zones, has been evaluated. Their stocking density has been standardized for cage production. The B:C ratio varied from 1.3 to 1.8 for table fish production. Recently the institute has developed and commercialized circular cage which can withstand high wind action and can be used for large scale fish production. The low cost CIFRI CageGrow feed developed by the institute is also supporting cage farming in reservoirs.

CIFRI has developed and introduced the CIFRI Pen HDPE, each unit covering an area of 0.1 ha, designed to securely contain fish and prawn stocks in open waters. This innovation serves the dual purpose of producing table-sized fish or raising stocking material (fingerlings). The pen features sides made of HDPE netting, with its bottom situated within the water body's bed. Utilizing pen culture for seed production or table fish cultivation significantly mitigates risks and threats associated with open water fish production. In particular, seed production through pen culture serves as an effective method for enhancing stocking initiatives, ensuring the timely availability of species-specific seeds in appropriate numbers and sizes. Extensive multi-location trials have been conducted with Indian Major Carps, minor carps, and minnows to validate the efficacy of the HDPE pen. The CIFRI pen culture technology has been embraced by several states in wetlands and reservoirs for both fish seed raising and table fish production, highlighting its versatility and applicability across diverse aquatic environments.

Climate resilience

Climate change poses significant threats to ecology and fisheries, impacting the aquatic environment, aquatic biota, fish diversity, breeding phenology, and fish production in inland open waters. The primary research focus of the institute has been on assessing the impact of climate change on ecology and fisheries, developing vulnerability assessment frameworks, understanding changes in fish breeding phenology, modeling fish reproduction and benchmarking reproductive traits, identifying climate-smart and sensitive species, evaluating the carbon sequestration potential of wetlands and greenhouse gas emissions, assessing ecosystem, species, and social vulnerability, investigating the dynamics of methanogens and methanotrophs, and identifying, documenting, and refining indigenous climate-smart fisheries adaptation strategies to mitigate extreme environments. Combining climate considerations into fisheries management is essential for promoting sustainable practices and maximizing the benefits of carbon sequestration. Identifying resilient species and integrating them into CBF regimes can enhance climate resilience. Enclosure farming in wetlands and reservoirs should prioritize species with low carbon footprints, improved feed conversion ratios, and high digestibility. Encouraging the cultivation of low-volume, high-value species and implementing carbon credit trading systems can incentivize eco-friendly practices, leading to both environmental preservation and economic growth. The institute has developed and demonstrated climate-resilient adaptation strategies such as Climate Resilient Pen System (CRPS) and Climate Resilient Culture-based Fisheries (CRCBF) in wetlands located in Assam, West Bengal, and Kerala. Climate awareness campaigns and educational programs are underway in these regions to enhance the adaptive capacity of fishers. The institute has also conducted studies on the impact assessment of recent natural calamities, including flash floods in Kerala, Fani in Odisha, and Yash & Amphan in West Bengal, on fisheries and the lives of fishermen. The findings from these studies have been published, and relevant advice has been provided to the concerned states and fishermen in the affected areas.

Natural farming

CBF constitutes a form of natural farming that taps into existing ecological dynamics. It involves the integration of multiple species, capitalizing on their complementary ecological niches to foster the growth of fish. Algae, phytoplankton, and zooplankton thrive with the support of organic matter and detritus, forming the foundation of the aquatic food chain. The production in CBF relies exclusively on natural food sources present within the water body. Striking a balance between ecological sustainability and maximizing fish yields remains an ongoing concern in natural fish farming. Modern approaches may incorporate some management strategies to optimize results while minimizing environmental impact in reservoirs and wetlands.

Empowerment of fisher communities

Empowerment of fishers is paramount for sustainable fisheries management and the well-being of communities. Providing fishers with access to education, training, and resources enhances their capacity to make informed decisions regarding fishing practices, fish farming, conservation efforts, and market opportunities. Empowerment initiatives also involve giving fishers a voice in policy-making processes, ensuring their perspectives are considered in regulations and management strategies. By empowering fishers, not only do we safeguard aquatic ecosystems and fish stocks for future generations, but we also foster resilient communities capable of adapting to environmental changes and economic challenges. Ultimately, the empowerment of fishers promotes the preservation of livelihoods, cultural heritage, and the aquatic environment. ICAR-CIFRI is instrumental in reaching to the unreached and providing inputs (fishing gears, crafts, fish seed, feed etc.), imparting skills through training and conducting various capacitybuilding programmes under STC, SCSP, NEH programmes and through other institutional and project activities.

Advances in open water fisheries management

Advances in inland open water fisheries management have been significant driven by technological innovations, scientific research, and adaptive governance approaches. One key advancement lies in the integration of data-driven decision-making processes facilitated by advancements in information and communication technologies. Remote sensing, GIS, and satellite imagery enable researchers and managers to monitor and assess aquatic ecosystems, identify fishing hotspots, and track changes in fish populations and habitats. Furthermore, the application of ecological modeling techniques, such as dynamic food web models and bioeconomic models, allows for the prediction of ecosystem responses to management interventions and the optimization of fisheries management strategies. Innovative fishing gear technologies including selective fishing gears and escape devices, have also been developed to minimize bycatch and reduce environmental impacts. Additionally, advances integrated multitrophic aquaculture systems and recirculatory aquaculture systems, provide opportunities for diversification. Moreover, there has been a growing emphasis on stakeholder engagement and participatory approaches in fisheries management, involving local communities, fisheries, scientists, and policymakers in decision-making processes. The integration of advanced technologies and data-driven approaches in inland open water fisheries management holds immense potential for enhancing sustainability efficiency, and resilience. Overall, these advances in inland open water fisheries management contribute to effective management and tapping the potential.

AI and Machine Learning Algorithms: These technologies enable the analysis of large datasets, prediction of fishery trends, and automation of decision-making processes. By improving the accuracy of stock assessments and resource management strategies, AI and ML algorithms contribute to sustainable fisheries management.

Data analytics and Sensors: Real-time monitoring of fish stocks environmental conditions, and fishing activities can be achieved through the use of sensors and data analytics. This information enables informed decision-making and facilitates the adaptation of CBF strategies based on up-to-date data. The institute has established a big data system encompassing biotic and abiotic parameters of major Indian rivers to facilitate sustainable and prudent management of the fisheries system through the application of data science and big data technologies. Furthermore, a Web-GIS application is in development to enable visualization of spatio-temporal data and generation of various scenarios. This interactive platform will offer a user-friendly interface for customized information browsing and report generation, enhancing the understanding of system dynamics efficiently.FISHPROT is a biological fish proteomics database for biomarker discovery and evaluation developed by the institute. ICAR-CIFRI is also developing sensor-based system for aquatic environment monitoring in collaboration with C-DAC, Kolkata.

Satellite technology (Remote Sensing & GIS): Satellite imagery and GIS technologies can provide valuable insights into open water conditions, resource mapping, and area delineation. They can also be used to monitor potential fishing zones (PFZs), identify priority habitats, track macrophyte coverage, and analyze fish migration patterns. Additionally, satellite technology can help detect illegal fishing activities, enabling more effective enforcement efforts. The institute has mapped waterbodies of different states of India in GIS platform (e-Atlas) which serve as support system for management decisions.

Drone technology: Drones offer a cost-effective means of collecting high-resolution imagery and data over large areas. They can be utilized for tasks such as environmental monitoring, resource mapping, and surveillance of fishing activities. By providing real-time data and imagery, drones enhance the understanding of the aquatic environment and support more targeted resource management efforts. The institute is a pioneer in using drone technology for water sampling and also mapping the macrophyte coverage in wetlands.

Mobile applications and Cloud-based platforms: Developing user-friendly mobile applications and cloud-based platforms for fishers, regulators, and other stakeholders facilitates data collection, access, and collaboration. These tools can enable fishers to report catch data, participate in collaborative management efforts, and access real-time information on fishery conditions and regulations. *eMATSYA* is an Electronic Data Acquisition System developed by CIFRI using the android application (in multilingual) to capture real-time fish catch data through mobile phones. The institute has also developed a sampling methodology and standalone software 'Catch Assessment Survey' (CAS) (Version 2.0) to estimate inland open water resources and fish catch production at the state-level.

Way forward

Realizing the untapped potential of reservoirs and wetlands holds the promise of boosting fish production by a staggering 4 mmt. Leveraging underutilized resources such as canals, coal pits, quarry ponds, check dams, upland lakes and minor irrigation tanks can further amplify the fish yields. Given that stocking plays a pivotal role in enhancement efforts, prioritizing the development of seed rearing infrastructure emerges as a critical necessity. Implementing stringent regulations to combat habitat loss, biodiversity decline, and ensuring habitat restoration is imperative. Moreover, there's a pressing need to focus on fostering climate resilience and adopting technologies with minimal environmental footprints. Integrating river ranching into fisheries development plans, guided by scientific principles, is essential. Establishing an integrated inland water management regime is crucial for planning and executing water resource projects, acknowledging both tangible and intangible benefits of aquatic resources. Policies must strike a delicate balance between sustainability and enhancement practices to achieve higher fish production, environmental integrity, and social equity. Co-management of inland water bodies is ideal, necessitating the active participation of all stakeholders in decision-making processes. Instituting effective tenure systems for fishers through policy, legislative, and institutional support is vital. Ensuring that fisheries have an equitable stake in open waters is paramount for safeguarding the interests of dependent fishers and fostering sustainable resource management.

Conclusion

Inland open water fisheries play a pivotal role in sustainably enhancing fish production to meet the growing demand for fish and to support the livelihoods of numerous marginalized fishers throughout India. There exists significant potential for boosting fish production in various inland open water bodies to maximize their fishery resources. Among the various strategies available, stock enhancement and culture-based fishery practices in wetlands and reservoirs stand out for their eco-friendliness and potential for long-term increases in fish production. The adoption of enclosure culture, utilizing technologies such as cages and pens, represents a promising approach for seed raising and table fish production, aimed at augmenting fish production from inland water bodies like reservoirs and wetlands. Furthermore, initiatives focused on fish diversity conservation, such as river ranching, habitat protection measures, implementation of closed seasons, pollution control efforts, and the assessment of environmental flow (e-flow), are crucial management options for riverine fisheries. By implementing these management strategies, stakeholders can work towards sustainable fishery practices that not only enhance production but also contribute to the conservation and preservation of aquatic ecosystems and biodiversity. Building capacity of stakeholders, creating value-chain and market links, building infrastructure and strengthening institutional and governance instruments are all essential for open water fisheries management. Advocacy for policies that prioritize sustainability and responsible management of open water fisheries, adaptable to evolving environmental conditions and scientific insights, is essential. Increased investment in research is necessary to foster innovative technologies and practices, enhancing the efficiency and sustainability of open water fisheries and thereby achieving SDGs.

Further reading

Bhattacharjya, B. K., Sarkar, U. K., Das, B. K., 2021. Fisheries Management modules for floodplain wetlands of India. ICAR-Central Inland Fisheries Research Institute, Barrackpore, p. 50.

Das, B.K., Sarkar, U.K., Samanta, S., Das, A.K., Hassan, M.A., 2021. Strategic plan for inland open water fisheries development under PMMSY. ICAR-CIFRI, Barrackpore. ISSN: 0970-616X

Das B. K., Chandra G., Meena D. K., Kumari S., Koushlesh S. K., Das A. K., Ekka A., Mishal P., Karnatak G., Bandopadhyay M. K., Samanta S., Pandit A., Sudheesan D., Manas H. M.,Mohanty B. P., Behera B. K., Sahoo A. K. & Parida P. K., 2017. Roadmap for Developmentof Inland Open Water Fisheries in Eastern States of India. ICAR-CIFRI, Barrackpore, pp 68. ISSN 0970-616X

Das B. K., Bhattacharjya B. K., Borah S., Das P., Debnath D., Yengkokpam S., Yadav A. K., Sharma N., Singh N.S., Pandit A., Ekka A., Mishal P., Karnatak G., Kakati A., Saud B. J., Das S. S. 2017. Roadmap for Development of Open Water Fisheries in North-Eastern States ofIndia. ICAR-CIFRI, pp. 120. ISSN 0970-616X

DoF, GoI, 2023. Annual Report 2022-23. Ministry of Fisheries, Animal Husbandry & Dairying, Govt. of India, New Delhi

FAO. 2024a. The State of World Fisheries and Aquaculture 2024. Blue Transformation in action. Rome, FAO. https://doi.org/10.4060/cd0683en accessed on 20.01.2025

FAO. 2024b. A review of the inland fisheries of India. FAO Fisheries and Aquaculture Circular, No. 1265. Rome. https://doi.org/10.4060/cd0674en

Sarkar, U.K., Mishal, P., Borah, S., Karnatak, G., Chandra, G., Kumari, S., Meena, D.K., Debnath, D., Yengkokpam, S., Das, P., DebRoy, P., Yadav, A.K., Aftabuddin, Md., Gogoi, P., Pandit, A., Bhattacharjya, B.K., Tayung, T., Lianthuamluaia, L. & Das, B.K. 2021. Status, potential, prospects, and issues of floodplain wetland fisheries in India: synthesis and review for sustainable management. Reviews in Fisheries Science & Aquaculture, 29(1), pp.1-32

Sarkar, U.K., Sandhya, K.M., Mishal, P., Karnatak, G., Lianthuamluaia, Kumari, S., Panikkar, P., Palaniswamy, R., Karthikeyan, M., Mol, S.S., Paul, T.T., Ramya, V. L., Rao, D. S. K.,Feroz Khan, M., Panda D.& Das, B.K.2018. Status, prospects, threats, and the way forward for sustainable management and enhancement of the tropical Indian reservoir fisheries: an overview. Reviews in Fisheries Science & Aquaculture, 26(2), pp.155-175

2

Present Status and Future Scope of Freshwater Aquaculture in India

Pramoda Kumar Sahoo**, S. Ferosekhan, Anirban Paul

Avinash R. Rasal and G.M. Siddaiah

ICAR-Central Institute of Freshwater Aquaculture, Kausalyaganga, Bhubaneswar Odisha

**Email: pksahoo1@hotmail.com*

Introduction

Aquaculture is one of the most promising and fast-growing food production sectors in the world with an estimated total fish production of about 178 million metric tonnes (MMT) (capture: 90.3 MMT; Culture: 87.5 MMT). The fish production from capture fisheries (90.3 MMT) has been almost stationary for many decades.However, the demand for fish is rising significantly due to the awareness about the health benefits of fish consumption (FAO, 2022). The additional demand for fish consumption must be achieved only through aquaculture. The fish production from aquaculture has increased from 44 MMT (2005) to 87.5 MMT (2020). Although the later sector is sharing almost 50 % of total global fish production, this current trend of growth is not sufficient to fulfil the future demand (FAO, 2022). The necessity for innovation and advanced technologies in aquaculture are highly needed for the expansion of the aquaculture industry.

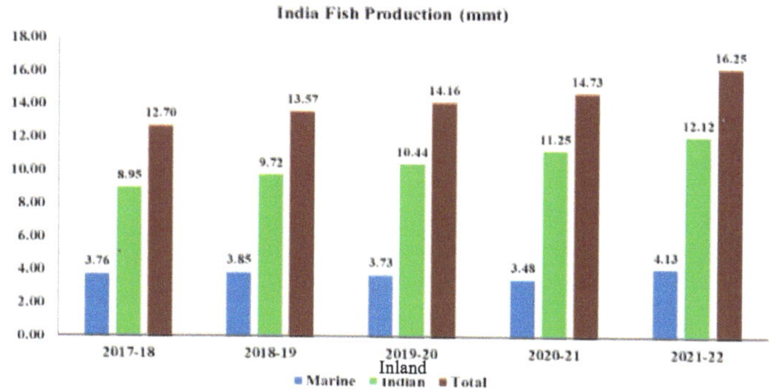

Fig. 2.1: Fish production trend of India

Status of Inland and Freshwater Aquaculture in India

India ranks first in the world on total inland capture fish production (1.80 MMT) followed by China (1.46 MMT) (FAO, 2022). During the last seven decades, Indian fisheries have grown more than thirteen folds with an increase in fish production from 0.75 MMT in 1950-51 to 16.25 MMT in 2021-22 (MoFAHD, 2023), of which the share from the aquaculture sector alone is around 7.79 MMT (FAO, 2022). Since the last three decades, freshwater aquaculture has also been compensating for the loss in fish production due to a decline in the capture fisheries as well as increasing overall fish availability in the country. Over the last four decades, the contribution of inland fish production to total fish production in India has registered a cumulative growth. It accounted for about 36% of total fish production in the 1980s, and this share increased to 75% by 2021-22 (MoFAHD, 2023). This increased growth rate is because of diversification and productivity increase in inland fish production besides horizontal expansion. In the past five years, aquaculture production has increased from 5.26 MMT to 7.79 MMT (FAO, 2022). In the year 2021-22, the top three inland fish production from Andhra Pradesh (4.22 MMT), West Bengal (1.65 MMT), and Uttar Pradesh (0.81 MMT). Similarly, the top fish producers were Andhra Pradesh (4.81 MMT), West Bengal (1.84 MMT) and Karnataka (1.07 MMT).

Andhra Pradesh dominates total and inland fish production in India, and it shares almost 30% and 34%, respectively.

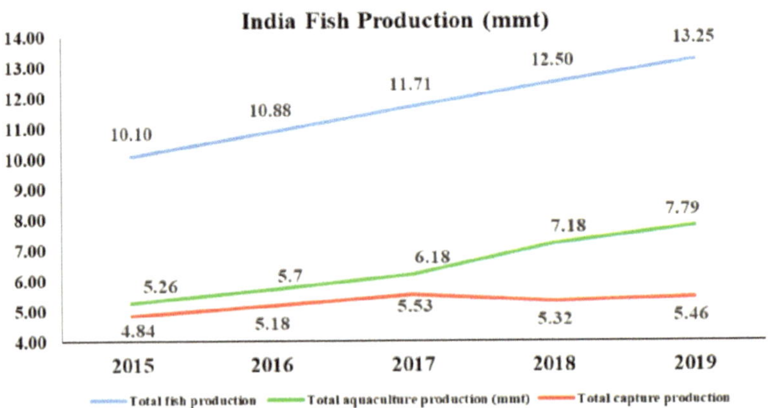

Fig. 2.2: Inland fish production trend in India

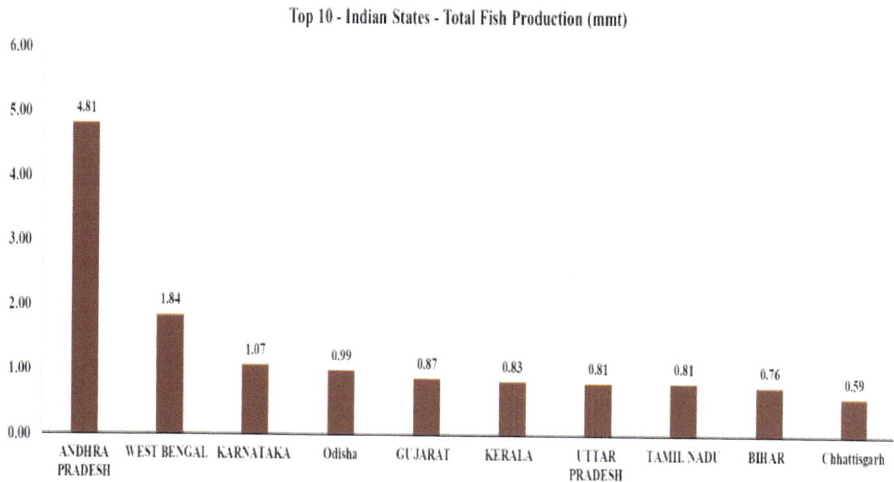

Fig. 2.3: Fish production by different Indian states

Status of Fish Seed Production in India

Fish hatcheries in both the public and private sectors have contributed towards the increase in seed production from 6322 million fry (1985-1986) to over 54069 million fry in 2020-21. West Bengal, Jharkhand and Assam are the top three major fish seed-producing states in India. The top 10 states contribute around 90% to the total seed production of the country (MoFAHD, 2023).

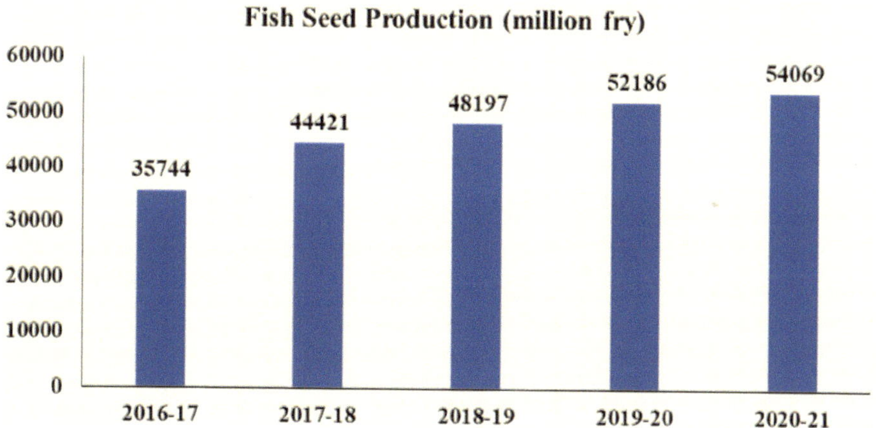

Fig. 2.4: Fish seed production trend of India

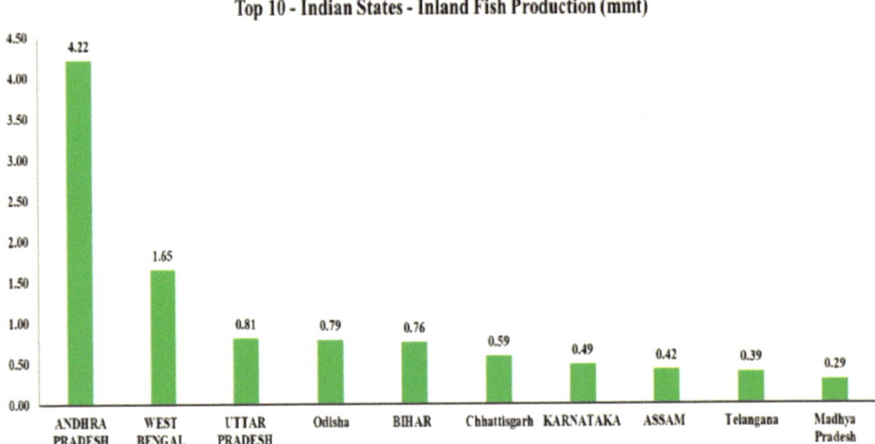

Fig. 2.5: Inland fish production by different Indian states

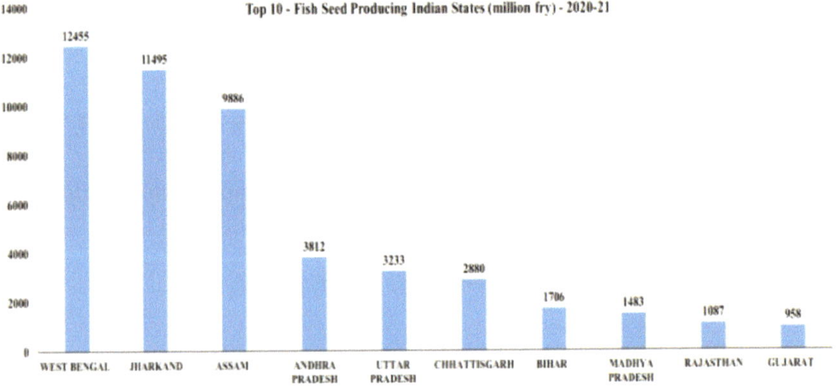

Fig. 2.6: Fish seed production by different Indian states in 2020-21

Priority areas identified in Indian Freshwater Aquaculture

ICAR-Central Institute of Freshwater Aquaculture, Bhubaneswar has been the nodal organization in India responsible for research, training, and extension in various aspects of freshwater aquaculture in the country. ICAR-CIFA is involved in excellence in research for developing sustainable and diversified freshwater aquaculture practices for enhanced productivity, quality, water use efficiency and farm income.

ICAR-CIFA has identified the following major research thrust areas to expand freshwater aquaculture production in the country.

- Aquaculture diversification (Species and System diversification)
- Genetic improvement for growth and disease resistance

- Fish nutrition and feed technologies
- Fish health management and advisories
- Socio-economic impact, policy research and ICT in aquaculture

Initiative by the Government of India (PMMSY)

Recognizing the vast potential for the advancement of the fisheries sector and emphasizing its importance, the Government unveiled a new initiative in its Union Budget for 2019-20 – the Pradhan Mantri Matsya Sampada Yojana (PMMSY). This visionary move aligns with the Hon'ble Prime Minister of India's multifaceted goals, which include doubling farmers' income, fostering self-reliance through "Atmanirbhar Bharat," embracing digital advancements in science and technology, and catalyzing a blue revolution within India. The introduction of the PMMSY scheme by the Indian government reflects its commitment to supporting rural communities, fostering entrepreneurship in aquaculture, enhancing fish production, encouraging greater consumption, and boosting exports in the fisheries sector.

Species diversification for freshwater aquaculture

India is blessed with rich fauna and flora diversity,and has diversified fish species and only countable species of them are being utilized for commercial aquaculture. Besides, Indian major carps (catla, rohu, mrigal), our country also possesses several other cultivable minor carp and barb species such as *Labeo calbasu*, *L. fimbriatus*, *L. gonius*, *L. bata*, *Systomus sarana* etc. Now-a-days, other fish species such as magur (*Clarias magur*), butter catfish (*Ompok pabda*), singhi (*Heteropneustes fossilis*), gangetic mystus (*Mystus cavasius*), pangas (*Pangasionodon hypophthalmus*), GIFT Tilapia, climbing perch (*Anabas testudineus*), murrels (*Channa striata* and *C. marulius*), pengba (*Osteobrama belangeri*) and freshwater prawn (*Macrobrachium rosenbergii*) are getting more attention for species diversification in Inland aquaculture sector. Further, the Institute has made breakthrough in developing breeding, seed production and culture technologies for peninsular carp species viz., *Puntius pulchellus*, *Hypselobarbus kolus*, *Labeo kontius*, *Barbodes carnaticus* which have great potential for the conservation of those species and also expanding Indian food fish culture basket.

Candidate species for Indian Freshwater Aquaculture Production

The following species are the key player in Indian freshwater aquaculture:

1. Indian major carps
2. Minor carps and barbs

3. Peninsular carps
4. Magur (*Clarias magur*)
5. Singhi (*Heteropneustes fossilis*)
6. Pabda (*Ompok bimaculatus*)
7. Striped catfish (*Pangasianodon hypophthalmus*)
8. Murrel (*Channa striata*)
9. Climbing perch (*Anabas testudineus*)
10. Ornamental fishes
11. Freshwater pearl production
12. Genetically improved rohu 'Jayanti'
13. Genetically improved catla
14. Genetically improved freshwater Prawn "*Macrobrachium rosenbergii*"

Indian Major Carps

Carps have been the mainstay of the Indian freshwater aquaculture & the majority of the Indian population prefers carp. Catla, rohu and mrigal are important carp species that are cultured traditionally in ponds and tanks due to their higher growth and consumer preference. Indian major carpproduction in India has increased from 3.59 MMT (2017) to 5.13 MMT (2021) and it shares almost 43% of total inland fish production of India during 2021 (FAO, 2022). Owing to the compatibility, with regards to food and habitat preference, these three species are cultured together in ponds under a polyculture system. Carp contributes about 80% of the Indian aquaculture production. Generally, in major carp polyculture, 6000-8000 fingerlings/ha are stocked for a production target of 4-5 tonnes/ha/year.

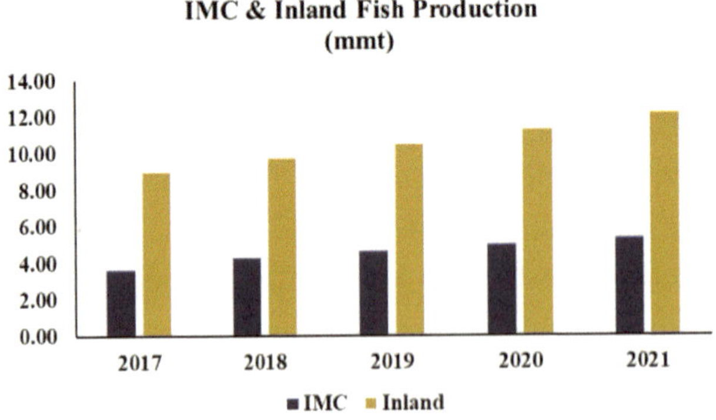

Fig. 2.6: Production Trends of IMCs in Relation to Total Inland Fish Production in India

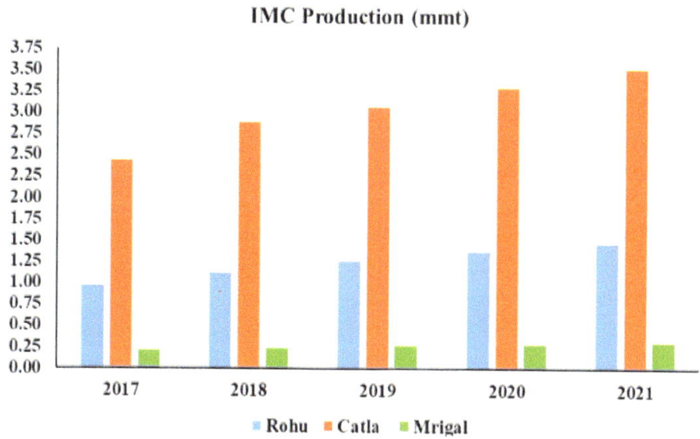

Fig. 2.7: Production Trends of the three species of IMCs

Minor carps and barbs

ICAR-CIFA has successfully standardised the breeding, seed production and culture of minor carps and barb species for species diversification. The minor carps and barbs such as *Lubeo fimbriatus, L. gonius, L. calbasu, Systomus sarana* and *Barbonymus gonionotus* come under this group. These species are recommended for polyculture with the Indian major carps. It was studied that polyculture of minor carps with major carps could increase production and generate additional income. Minor carps and barbs fetch a good market price in different regions. These are greatly recommended for species diversification. Incorporation of minor carps with IMC leads to 20-25% higher yield and 30% higher profit through minor carp intercropping with carp polyculture.

Magur

Magur, *Clarias magur* is a potential catfish species for inland aquaculture production. The culture of magur is mostly done by monoculture. Assam, Odisha, Bihar, West Bengal and a few eastern and northeastern states are major magur-producing states. It has very good market demand and is considered ahigh-value catfish (Rs.300-600/kg). The culture practices of magur are being undertaken in small earthen ponds. Generally, a high density of 40,000-60,000/ ha is recommended for grow-out culture. Bigger-sized seed (3-5g) shows good survival and growth in grow-out culture phase. The fishes attain a marketable size of 100-150g during a period of 10 months of culture. Harvesting is done by completely dewatering and picking them manually from the culture ponds. Average productions of 3-4 tonnes/ha can be achieved. ICAR-CIFA has successfully standardized the breeding, seed production and grow-out culture of magur and actively involved in technology dissemination.

Singhi

Singhi or stinging catfish (*Heteropneustes fossilis*) is an excellent candidate species for species diversification. It fetches good consumer preference and a market price of Rs. 300-500 per kg. ICAR-CIFA has standardized the breeding and seed production techniques of *H. fossilis*. The breeding season of *H. fossilis* is normally between June to August, and it responds well to captive spawning, and male need not be sacrificed for undertaking breeding. The seed-rearing and grow-out culture is very similar to magur.

Pabda

ICAR-CIFA has standardized the breeding, seed production and grow-out culture protocol for pabda. It attains maturity at the end of the first year. It matures at the size 40-60 g weight. It is a monsoon breeder. The grow-out culture of pabda is generally carried out with stocking fingerlings at 25000-30000 nos./ha and the production of 1-2 tonnes/ha can be achieved.

Striped catfish

Pangasianodon hypophthalmus was first introduced into India from Bangladesh in 1997 in West Bengal. Thereafter in 2004, the catfish species was introduced in Andhra Pradesh where massive production was achieved. Due to higher productivity and faster growth rate, striped catfish culture spreaded to many other states. Pre-stocking management of pangasius is similar to that of IMC culture. Farmers stock seeds of 50-150g at 20,000-25,000 fingerlings/ha. Grow-out culture takes place for about 6-8 months and reaches 0.75 to 1 kg with the production of 15-20 tonnes/ha. ICAR-CIFA has standardized the breeding, seed production and grow-out culture of striped catfish.

Striped murrel

Murrels are referred as snakehead due to their elongated and cylindrical body, and flattened head. *Channa striatus* is a highly carnivorous fish and requires a high protein diet. Striped murrel fetch high market price and is having huge demand in southern states. The stocking density for grow-out under monoculture is 10,000-20,000 fingerlings/ha of the size >10 g. It reaches 0.5-0.75 kg in 10 months culture period with the production of 2 tonnes/ha. ICAR-CIFA has standardized the breeding, seed rearing and grow-out culture of murrels. Technology transfer has been made to farmers across the country and to the Govt. of Telangana and Tamil Nadu for wider dissemination of the technology.

Climbing perch

Climbing perch (*Anabas testudineus*) is popularly known koi and it has good market demand in West Bengal, Tripura, Assam, Manipur, Jharkhand, Bihar and Kerala. Climbing perch is a potential culture species with high economic value and preference. Due to its air-breathing habit and tolerance to adverse environmental conditions, this fish turns out to be an ideal candidate species for culture in the derelict water bodies.The stocking density for grow-out culture is 4-6 fingerlings/m^2. The fish attains marketable size at 50-80g in 6 months. For harvesting, the pond must be dewatered completely, and fish are hand-picked. The production level of 2-3 tonnes/ha/6 months can be achieved. ICAR-CIFA, Bhubaneswar has standardized the breeding, seed production and culture technologies of this species.

Breeding and culture techniques of popular ornamental fishes

The indigenous and exotic freshwater species of ornamental fish varieties having good demand, and which can be bred and reared for commercial purposes are broadly grouped into two categories on the basis of breeding behavior *i.e.* livebearers and egg layers. Livebearers are the ones who bear live fish. Egg layers are of the normal egg-laying varieties like most other fishes.

Shining barb

For the first time ICAR-CIFA has developed and commercialized a new variety of rosy barb (*Pethia conchonius*) called "Shining barb", through selective breeding. The developed variety is more colourful than its parents. There is golden shining on the head and dorsal side of the shining barb. A glittering gold colour in females and shining pink-red in males is expected to fetch a significantly higher price than its normal counterpart available in the aquarium industry.

Breeding of barbs

Broodstock of barbs is to be separated sex-wise one month before breeding with the feeding of balanced diet and one-time live food. Breeding nests are provided using plastic filaments in breeding tanks with one foot of water level, where breeding pairs of 1:1 or 2:1 (male: female) are maintained. After spawning, broods are to be removed from the tanks and proper post-breeding care has to be given to them. The sticky golden-coloured eggs are found attached to nylon filaments and these are allowed to hatch in the same tank. After three days of spawning, larvae are to be fed infusorians or small rotifers for further development for a period of one week. After a week they may become acclimatized with powdered feed and zooplanktons like daphnia and moina for about 15 days. After a fortnight, they may be shifted to culture tanks for further rearing. They reach marketable size within a period of three to four months.

Freshwater Pearl Production

Pearl often referred as the "queen of gems" having rich cultural significance, exhibits tremendous market demand and in recent times emerged as the most lucrative aquaculture enterprise in countries having natural mussel resources. In old days, the pearl was collected from wild sources, but in recent days the technological advancements in pearl culture have been studied and standardized, which makes the way to develop pearl culture technologies in molluscan bivalves of freshwater and marine origin. The Institute is working towards the development of freshwater pearl mussel culture technologies of *Lamellidens* species and dissemination of freshwater mussel pearl production technology to the various stakeholders, farmers, and entrepreneurs through training programmes. Pearl mussels of 35-50 g weight and 8-10 cm long are selected to carry out surgical implantations. Prior to the surgery, the collected mussels are crowded for 24 to 48 hours to ensure that the adductors are relaxed for the ease of nucleus implantation. The pearl formation takes place around 10 months period and the culture can be carried out in glass tanks, cement tanks and ponds. Freshwater pearl production is becoming a lucrative business and it has huge international market demands.

System diversifications

Recirculating Aquaculture Systems (RAS)

Recirculating Aquaculture System (RAS) represents a novel and innovative approach for fish farming within a controlled environment. This method aims to optimize production while minimizing spatial requirements, enhancing water usage efficiency, enabling higher stocking densities, reducing labour demands, facilitating fish feeding, grading, and harvesting, and, notably, water filtration and reusing water without any exchange. RAS is swiftly gaining recognition and widespread acceptance as a sustainable approach to the cultivation and responsible management of aquatic species. RAS employs multiple filtration stages to treat water, allowing its reuse within the same system. The typical combination of stages includes mechanical filtration, biological filtration, and UV and ozone sterilization. In this process, mechanical filtration primarily eliminates solid waste, while biological filtration effectively removes dissolved waste and helps ammonia conversion. Subsequently, sterilization steps work to decrease bacterial and pathogen concentrations throughout the entire system. ICAR-CIFA has established RAS facility for catfish breeding and seed production and also for murrels brooder raising and hi-density seed rearing for mass scale production.

Biofloc technology

Biofloc is defined as macro aggregates composed of diatoms, macro algae, fecal pellets, exoskeleton, remains of dead organisms, bacteria, and invertebrates.In this system, a co-culture of heterotrophic bacteria and algae is grown in flocs under controlled conditions within the culture pond. If carbon and nitrogen are well balanced in the solution, ammonium in addition to organic nitrogenous waste will be converted into bacterial biomass. Nitrogen control in biofloc is induced by adding carbohydrates into the system, and subsequent uptake of nitrogen by heterotrophic bacteria which resulted in the synthesis of microbial proteins through assimilation thereby removing ammonium from the water. Some of the major benefits the technology offers to the culture practices are improvement of water quality, serving as a nutrient for fish, acting as an antimicrobial agent and increasing fish production. Biofloc technology provides high productivity, low feed-conversion ratios and a stable culture environment. Biofloc technology can help deliver sustainable production at a lower cost. ICAR-CIFA has established a biofloc research facility and standardizing the seed production and grow-out culture techniques of high value fishes such as singhi, magur, anabas, scampi, carps and barbs.

Aquaponics

Aquaponics, the integration of the Recirculatory Aquaculture System (RAS) and hydroponics (soilless horticulture) is getting overall consideration from researchers and fish culturists as a result of its high ecological and monetary advantages. The Nutrient Film Technique (NFT) aquaponic system has been designed and developed under AICRP at Plastic Engineering in Agriculture Structure and Environment Management (PEASEM) Centre at ICAR-CIFA, Bhubaneswar. The aquaponics unit consists of four major components, such as a Fibreglass Reinforced Plastic (FRP) round fish culture tank with operational capacity 2800 L, a biofilter unit made up of Polypropylene (PP) of 100 L capacity, FRP rectangular hydroponics tank having 2.64 m^2 plantation area with 84 perforations for holding plantation net cups and High-density Polyethylene (HDPE) sump of 200 L capacity. Fibreglass Reinforced Plastic (FRP) composite was chosen as the primary material as it is having higher yield tensile strength of about 13% than steel with significantly lesser weight. Also, thermal insulation, UV resistance, corrosion resistance, ease of fabrication, resistance to chemical erosion, less maintenance and easy transportation in comparison to cement or steel were key factors for selecting FRP as the material. Implementation of custom designed and calibrated automatic water recirculation system gives an average flow rate of 94.7 L/h for continuous flow of nutrients from the fish culture tank to hydroponics tank. The designed system harnesses gravity flow in 75% of the cycle. The designed biofilter showed an average 61.97% reduction in ammonia with a hydraulic loading of 3.68 LPM (liters per minute)/m^2 of filter media.

Freshwater Integrated Multi-tropic Aquaculture (FIMTA)

FIMTA is the farming of aquaculture species of different trophic levels in proximity with complementary ecosystem functions, in a way that allows uneaten feed, waste, and by-products of one species to be utilized as fertilizers, feed, and energy for the other crops, and to take advantage of synergistic interactions among the species. IMTA facilitates biomitigation, and diversification of fed monoculture practices, by combining them with extractive aquaculture species, to realize benefits environmentally, economically, and socially.

Portable Hatchery System

ICAR-CIFA has developed technologies such as FRP Portable Carp Hatchery, FRP Magur hatchery and FRP Padba hatchery.

FRP Portable Carp Hatchery

Fiberglass Reinforced Plastic (FRP) carp hatchery has proved to be a very effective tool in carp seed production which will be beneficial to the farmers. It can be transported, installed and operated in remote places to ensure easy and timely availability of carp seeds. FRP hatchery is suitable for fish breeding in field conditions for 10-12 kg of carps in one operation and can be used as a tool for bio-diversity conservation. The portable hatchery is a complete set of plastic structures for the spawn or hatchling production from matured fishes of carps. The FRP carp hatchery operates with the principle of eco-hatchery. One complete unit consists of (i) Breeding/ spawning pool, (ii) Hatching/ incubation pool, (iii) Egg/ spawn collection chamber, and (iv) Overhead storage tank and water supply lines. It is suitable for small-scale breeding with a production capacity of 1.0-1.2 million spawn of Asiatic carps in one operation. Its efficiency is cent percent in field conditions. With spawn availability in one operation from the hatchery unit and with proper seed-rearing practices, about 30 hectare ponds can be stocked with good quality fish seed. The profit to the hatchery operator can be 2-3 lakhs rupees in one year with full operation of the hatchery unit. At present 27 states have installed hatcheries and around 210 users were direct beneficiaries of the technologies. It is developed under ICAR-AICRP on PEASEM, Centre at ICAR-CIFA, Bhubaneswar.

FRP magur hatchery

The portable magur hatchery is a simple device comprising a stand on which are placed a row of plastic tubs (12 cm dia, 6 cm high). Water is supplied from the overhead tank through a common pipe to all the tubs with individual control tabs. It includes egg incubation and hatching. The technology creates a suitable environment for a high hatching percentage where a maximum of 50,000 eggs can be incubated at a time.

FRP pabda hatchery

A multi-tier hatchery system for pabda seed production has been developed for the first time. The hatchery unit made up of FRP consists of three major parts i.e., Hatching/incubation unit; brood acclimation – cum – latency stage nurture cum hatchling rearing tanks and overhead storage tank/water supply system. The cylindro-vertical hatching/incubation unit placed one above the other on a triangular iron frame is a unique design that not only minimizes the requirements of space but also gives operational ease. This unit can accommodate about 40,000-50,000 nos. of eggs in a single cycle. Three rectangular tanks of different sizes are designed for multiple purposes viz., for acclimation of pabda broodfish after injecting inducing hormone and for the rearing of hatchlings to the early fry stage. Using these rearing tanks 10,000-15,000 nos. early fry can be produced in a single cycle.

Genetic Improvement Program for Rohu, Catla and Scampi in India

Selective breeding in aquaculture is a long-term process that aims to genetically improve fish species for economically important traits such as growth rate, disease resistance, feed conversion efficiency, and reproductive fitness. This method involves carefully choosing individuals based on their genetic parameters and pedigree information. Over the past five decades, selective breeding in aquatic species has shown remarkable success in achieving genetic improvement using family/individual selection, resulting in a significant 12.5% (average) increase in growth rate per generation.

Rohu Selective Breeding Program

Rohu, *L. rohita* has become well-known as a promising aquaculture species in nations including India, Bangladesh, Myanmar, and Pakistan. Due to effective induced breeding and domestication of this species, India and Bangladesh have used crossbreeding and selective breeding methods for genetic improvement. The ICAR-Central Institute of Freshwater Aquaculture (CIFA) in Bhubaneswar, Odisha, India, and the AKVAFORSK (Institute of Aquaculture Research) Norway jointly launched the selective breeding program for rohu in India in 1992. Gjerde *et al.* (2002) performed two complete diallele crosses with five stocks of Rohu carp (*Labeo rohita*) to find the effect of heterosis on growth and survival and reported that for harvest weight and survival, total heterosis for each of the stock crosses was low or negative. The process involved establishing a base population using six stocks collected from various rivers, including the Ganga, Yamuna, Sutlej, Gomati, and Brahmaputra, along with a farmed stock of CIFA. Full-sib families were generated using the diallel-crossing, and the fingerlings were tagged with a passive integrated transponder (PIT) for the evaluation of genetic parameters. Gjerde *et al.* (2019) studied the genetic parameters for growth and survival in rohu breeding program from data recorded on 16,718 rohu carp (*Labeo rohita*), the offspring of 311 sires and 257 dams from seven year-classes. The estimated heritabilities (and of the effect common to full-sibs, c^2) across the two production systems were 0.22 ± 0.15 (0.66 ± 0.07), 0.38 ± 0.11 (0.28 ± 0.05), 0.34 ± 0.10 (0.23 ± 0.04) and 0.14 ± 0.05 (0.08 ± 0.02) for body weight at tagging, at sampling, at harvest and survival until harvest (on liability scale), respectively.

Genetic improvement was carried out through a combined selection method. Over eight generations, an average genetic gain of 18% was achieved in rohu, resulting in the development of a variety called "Jayanti" rohu. The selective breeding program has continued, with 13 generations completed to date, and each generation consists of 50-60 families in each year class. Furthermore, a disease resistance trait against the bacterial pathogen *Aeromonas hydrophila* was incorporated into the rohu selective breeding program, where arrays of single nucleotide polymorphisms (SNPs) were identified using transcriptome data. Estimated heritabilities for challenge test survival obtained from the threshold model in Jayanti rohu is 0.11±0.04 (Mahapatra *et al.*, 2008). Experimental infection with *Aeromonas hydrophila* (9.55 X 10^6cfu/ g fish) through the intraperitoneal route produced higher survival in the resistant line (73.33%) as compared to the susceptible line (16.67%) (Sahoo *et al.*, 2011). The resistant line showed 56% greater survival in the challenge test over the susceptible line. ICAR-CIFA received a Trade Mark Registration for "AhR Jayanti" rohu (*Aeromonas hydrophila* resistant Jayanti rohu) in April 2023.

A systematic study with sixth generation Jayanti rohu under normal aquaculture condition performed by NGO, Kalong Kapili, Assam reported that Jayanti rohu grows to 84% higher (1.2 kg in a year) as compared to normal rohu of (0.65 kg in a year) (Mahapatra *et al.,* 2016). Ingtipi et al.(2021) studied the growth performance of 9th generation Jayanti rohu with non-Jayanti rohu under polyculture system with *Catla catla* and *Cirrhinus mrigala* at different stocking densities for 90 days and found that the Jayanti rohu grows 6.65 to 31.6% higher compared to non-Jayanti rohu in Assam. Saikia *et al.* (2020)

reported that in 10 months grow out culture from fingerling size (10-15 cm), Jayanti Rohu showed a significantly higher average growth of 740±28.1 g (31.6 % higher) compared to local rohu which grows average 562.17±16.85 g in three locations viz., Chotobinyakhata, Hatigarh and Dhauliguri from Kokrajhar district of Assam. The Jayanti rohu was demonstrated in several parts of the country; however, the states like Odisha, West Bengal, Assam and Andhra Pradesh are the major areas of adoption in the country.

Catla Selective Breeding Program

Catla is the second most important IMC after rohu and is immensely popular due to certain characteristics such as fast growth, taste and market demand. Catla is indigenous to major riverine systems of India, Pakistan and Bangladesh and is of high worth for aquaculture production in India and the whole Indian subcontinent.ICAR-Central Institute of Freshwater Aquaculture (ICAR-CIFA) has initiated a selective breeding programme on Catla (*Labeo catla*) to improve the body weight at harvest and quality seed production during 2010 to cater to the needs of the farming community. Nine strains/populations of *L. catla* including two riverine strains (Ganga and Subernarekha) from different geographical regions of the country viz. West Bengal, Bihar, Odisha, Andhra Pradesh and Uttar Pradesh were collected for the establishment of the base population (Mahapatra et al., 2018). Combined family selection method was used for selecting superior animals for every generation based on their breeding value (Mahapatra et al., 2016). Ideal size of catla fingerling for PIT tagging is 20-30 g (Rasal et al., 2022). Till date three generations of improved catla are produced. After two generations of selection, 15% genetic gain per generation was obtained in genetically improved catla (Mahapatra et al., 2022). Under the field trials in different states (Odisha, West Bengal, Assam and Maharashtra), the genetically improved catla, in the polyculture system, have attained a mean weight of 1.8 kg in comparison to the local strain of 1.2 kg, across all locations in one year when size at stocking was 50 g (Mahapatra *et al.*, 2022).

Dissemination of Genetically Improved rohu (Jayanti) and Improved catla from Nucleus Breeding Centre

The ICAR-CIFA is the sole NBC for Jayanti rohu improved catla selective breeding program in India. The Jayanti rohu and improved catla are disseminated in a three-tier structure where ICAR-CIFA provides brood seed to multiplier units to raise the broodstock and followed by mass scale production of seed of improved varieties to distribute to farmers across the country for grow-out culture. During the decade from 2010 to 2021, a total of 215 million spawn of Jayanti rohu were disseminated to 18 different states/ union territories. Similarly, during the 2014 to 2023, a total of 20 million spawn of improved catla have been disseminated to 11 different states/union territories. For dissemination of the Jayanti rohu and Improved catla among various stakeholders, ICAR-CIFA signed Memorandum of Understanding (MoU) with seven commercial hatcheries/institutions from Odisha, Assam, Andhra Pradesh, Tamil Nadu and Maharashtra and since 2016, National Freshwater Fish Brood Bank (NFFBB) of the National Fisheries Development Board (NFDB), Hyderabad is a major disseminator of the technology to the hatchery operators in our country.

Aeromoniasis resistant rohu

Aeromoniasis is the disease caused by *Aeromonas hydrophila* that have a major impact on carp production in India especially in Andhra Pradesh. By improving the disease resistance of aquaculture stock, India could reduce the cost of production, positively affect both the local environment and human health by reducing the use of antibiotics and other chemical treatments and gain a major advantage in the production of these important food species. Selection for disease resistance in fish may be performed directly on basis of survival data obtained in controlled challenge trials, or indirectly using information from immunological or molecular markers linked to differential survival.

Disease resistance against *Aeromonas hydrophila* is added as a second trait of selection to the rohu breeding program. Resistant line rohu showed significantly high survival over the susceptible line. Estimated heritability for challenge test survival obtained from the threshold model in Jayanti rohu was 0.11±0.04. The first generation of resistant line showed a higher survival (56.67%) over the susceptible line in the challenge test. This result showed clearly the inheritance of the resistance trait in genetic lines of rohu which may give the possibility of increased disease resistance on a long-term basis. During the year 2006, offspring were produced from susceptible and resistant families. Two extreme lines (resistant and susceptible) were produced. The offspring of both the lines were challenged and the realized response to selection was found to be 31.1% in first generation itself and in the second generation it was 58% (Mahapatra *et al.*, 2016). Further positive correlation (0.43) was observed between growth and disease resistant trait i.e. against *A.hydrophila* in the rohu breeding program.

ICAR-CIFA received the trademark for disease resistant rohu against *Aeromonas hydrophila* (AhR Jayanti). ICAR, New Delhi has certified the technology of AhR G-13 Jayanti rohu on the eve of ICAR foundation day 16 July 2023.

Genetically Improved Freshwater Prawn (CIFA-GI SCAMPI™)

Giant freshwater prawn, *Macrobrachium rosenbergii* is an important freshwater crustacean widely cultured in several countries including India. Of late, its production has come down due to a slow growth rate and disease occurrences. The Institute in collaboration with WorldFish Center, Malaysia started a genetic improvement program of *M. rosenbergii* through selective breeding in 2007. For establishment of the base population, stocks of *M. rosenbergii* were collected from three geographically distant locations (Gujarat, Kerala and Odisha) and stocked at ICAR-CIFA. CIFA-GI SCAMPI™ is the first genetically improved prawn in India. It has shown 7% improvement in weight gain per generation. ICAR-CIFA has developed a genetically improved variety of giant freshwater prawn, registered as CIFA-GI Scampi in 2020, with a 53% higher growth rate, 68% higher yield, fetched 13% higher farm gate price with Rs. 0.62 lakhs/ha/crop additional income compared to non-GI scampi in carp-scampi polyculture system. Multi-location performance trials of GI scampi in carp-scampi polyculture system were evaluated in farmers' ponds in Odisha, West Bengal, and AP during 2021 & 2022. ICAR-CIFA has signed MoU with five scampi hatcheries in AP and also with the National Freshwater Fish Brood Bank (NFFBB) as the multiplier units of CIFA-GI SCAMPI® for wider dissemination of improved scampi seeds.

Feed Nutrition and Feed Technologies

Freshwater aquaculture in India is an important economic activity and fastest growing animal protein producing sector with an annual compound growth rate of over 8%. The vitality of the freshwater aquaculture can be witnessed by the 20-fold increase in fish production from 0.37 million metric tons (MMT) in 1980 to almost 7.6 MMT in 2022; surpassing capture-based fish production. It specifies that aquaculture would be mainstay for providing the majority of table fish for human consumption. To maintain current growth rate, nutrients and feed inputs has to grow at a consistent rate. Feed is the most important farming input, accounting for 50–70% of overall production costs. Currently, the majority of carp farmers use traditional bag feeding methods that include a blend of rice bran and oil cakes (economic FCR of 1:2.5 -3.0) as well as apply fertilizer to maintain sufficient pond primary productivity. It is estimated that 9-10 MMT of farm-made feed are used in carp culture by small and marginal farmers. Certain carp producers employ commercially supplied floating feeds (10-12%), and the use of commercial feed is determined by the farm gate price of the fish. On the other hand, approximately 1.1 MMT of commercial feed are exclusively used by Pangasius, Tilapia and freshwater prawn growers.

To achieve a target fish production from freshwater aquaculture of 13.5 million metric tons, an estimated 20.25 MMT of commercial feed is required (assuming an FCR of 1:1.5). If semi-intensive conventional feeding practices (mash feed or farm-made feed) in carp culture continue in the future, the feed requirement will be 40 MMT with an economic FCR of 3.0. To ensure the long-term sustainability of the aquaculture sector, feed-based culture must be increased by at least 40–55%. For this, the Government of India has made significant efforts through PMMSY's many schemes, most notably the approval of an additional 465 commercial feed mills/plants across the country. The constraints faced by the aquaculture industry are viz., i) increased feed cost due to price volatility of raw materials inadequate supply and competition from other food-producing sectors, such as poultry, piggery, and dairy for a same ingredient (soybean meal, deoiled rice bran, and other oil cakes), ii) higher FCR of existing freshwater aquaculture feeds and high load of solid waste into production system, and iii) lack of species specific larval and brood stock diets for diversified species. Therefore, the future scope of research and development in Fish Nutrition and Feed technology would be

a) ***Search for alternative feed resources and processing technology:*** The government of India's effort for bioethanol generation for fuel is expected to result in the growth of ethanol-producing industries. These industries produce a large number of by-products, specifically dry distillery grain soluble (DDGS), with outstanding nutritional properties. Other by-products from the brewery sector include brewery-spent grain (BSG) and brewery-spent yeast (BSY). Again, a vast number of by-products from the poultry, meat, and food processing industries can be used as fish feed ingredients, partially replacing expensive feed materials. There are lot of underutilized plant proteins and by-products from agro-industry as fish feed component due to presence of antinutrients. Hence, there is an urgent need for novel approach or technology to partially or completely remove the antinutrients while increasing the nutritional content of these components. Database on maximum dietary inclusion levels of all locally available ingredients (regional basis) and their digestibility must be studied and prepared which would help fish producers and feed manufacturers.

b) ***Improving feed efficiency:*** With this limited feed resources and higher FCR of existing feeds available in India, it is necessary to improve the FCR. One of the UN Sustainable Development Goals is to reduce the carbon footprint of the food-producing industry. It is noted that feed is a significant contributor to greenhouse gas emissions in aquaculture.

Therefore, reducing emissions requires improved feed efficiency (nitrogen and phosphorus retention in the body) and minimizing solid waste output into culture system. This can be accomplished through improved feed formulation (based on amino acid requirements rather than crude protein and identifying species-specific cost effective suitable functional feed additives) and effective feeding management approaches.

c) ***Larval inert diets:*** It is expected that increased demand for fish seed and a lack of brooders in the future will necessitate enhancements in larval survival. Given the high expense of live feed production, additional research and development into the formulation of high-performance larval inert diets is urgently required.

d) ***Broodstock diets:*** As maternal nutrition influences the health and well-being of offspring, identifying critical nutrients required for optimal reproductive performance is essential part of research.

e) ***Tailored feeds:*** The Biofloc system, RAS, and in-pond raceway are expected to have significant potential for increasing fish production in India. Feeds formulated specifically for these husbandry procedures are required. It is also vital to formulate a finisher diet which improves the product quality (enriched nutrients of human health benefits). In this regard, ICAR-CIFA has been instrumental in research and development for enhancing freshwater fish production. The developed and commercialized feed technologies are as follows:

CIFA-CARP STARTER

Nursery phase is one of the most critical phases in the production of quality fingerlings for carp culture. Generally, it is observed that the recovery of spawn to fry and fry to fingerling are low (25-30% and 40-50%, respectively in farmers' practice). To address this problem, ICAR-CIFA, Bhubaneswar has developed a carp nursery feed which ensures over 80% survival, better growth and uniform-sized fingerling production. This feed is highly nutritious and palatable. The feed is suitable for carp seed growers to enhance their production and profitability.

CIFA-CARP GROWER

The carp grower phase is one of the most important phases for production of marketable fish in carp culture. Generally, it is observed that the profit from using commercial/handmade feed in grow-out culture is very minimal, and the taste of the fish is also not up to the satisfaction of customers. The fish may not be healthy by using commercial feed. To address this problem, ICAR-CIFA,

Bhubaneswar has developed a carp grower feed which ensures better growth, improved FCR, higher palatability to fish, disease resistance and tastier fish. This is suitable for carp growers to enhance their production and profitability.

CIFABROOD™

CIFABROOD™ is an exclusive carp broodstock diet, adequately rich in essential nutrients. It advances gonad growth and maturation, facilitates early spawning and significantly increases spawning response. It is suitable for multiple/ repeated breeding, off-season gonad growth and post-spawning recovery. The rate of feeding is 3-5% of total body weight during the egg development phase. Pre-monsoon breeding was carried out in CIFABROOD™ demonstration ponds in various states including West Bengal, Assam, Tripura and Bihar. Monsoon breeding was carried out in states including West Bengal, Assam, Tripura, Chhattisgarh, Odisha, Tamil Nadu and Bihar. Breeding and seed production of IMCs during North-East Monsoon was carried out at Pugazh Aqua Farm, Cuddalore (on 3rd December 2019) producing 25 lakhs of spawn and Govt. Fish Farm, Manimuthar, Tamilnadu (on 5th, 9th, 12th December 2019 and 9th January 2020) producing 40.0 lakhs of spawn, which is the first time in India and regarded as a breakthrough. Initial impact analysis suggests CIFABROOD™ has increased the efficiency of the brood stock by raising the productivity of the female by 147%. On the other hand, it is able to reduce the net cost of Rs 120 per lakh of spawn which is about 40% of the cost of production.

Fish Health Research in Freshwater Aquaculture

Several developments have taken place for effective management of fish diseases in freshwater aquaculture. The Institute has been a frontrunner in the context of disease diagnostics and management of freshwater fish diseases. ICAR-CIFA initiated a trial of ribosomal P0 peptide as a potential vaccine candidate for control of fish ectoparasite infections, i.e., *A. siamensis* in *L. rohita* and reported delayed mortality with relative percent survival of 16% following experimental parasite challenge (Kar *et al.*, 2017). In addition, several recombinant proteins of *Aeromonas hydrophila* were experimented for the development of an effective vaccine against *A. hydrophila*. Recombinant outer membrane protein R (rOmpR)-based vaccine of *A. hydrophila* with a modified adjuvant formulation in rohu has shown great potential for disease management with elevated immune status of fish (Dash *et al.*, 2014). Recombinant outer-membrane protein F (rOmpF) of *A. hydrophila* significantly increased the relative percentage survival (~44%) in the vaccinated group in *Labeo rohita* (Yadav *et al.*, 2018). Further, recombinant outer-membrane protein C (rOmpC) of *A. hydrophila* significantly increased the relative percentage survival (~77%) in the vaccinated group which is significantly higher than rOmpF (Yadav *et al.*, 2021). Recently, ICAR-CIFA signed MoU with Indian Immunologicals Limited for commercial production of *A. hydrophila* vaccine candidate developed by CIFA. Further, under a DBT supported project, the institute is developing a TiLV vaccine to protect tilapia against this virus.

Bacterial diseases are the most common form of disease in the Indian aquaculture scenario. Among different bacterial infections, *Aeromonas hydrophila* mediated infection is the most frequent one. To combat this ICAR-CIFA has produced AhR-JAYANTI rohu which is a disease-resistant variety of rohu against bacteria *Aeromonas hydrophila* developed through selective breeding.

ICAR-CIFA has also developed several fish health management products which directly benefits farmers by reducing disease associated loss. Among several products developed, *CIFAX* is a chemical formulation and first commercialized technology of CIFA. It prevents and cures ulcerative diseases (EUS) of freshwater fishes and is widely used by the farmer's community. Another formula *i.e.CIFACURE* is used for controlling common bacterial and fungal infections of freshwater ornamental fishes. The product can be very well used in aquariums and other outdoor tanks where the ornamental fishes are grown. It controls many bacterial diseases like haemorrhagic septicemia, ulcers, fin rot, tail rot, eye diseases and mouth fungus and other fungal infections. Recently, to treat the external injury associated with argulosis and other secondary bacterial infection, CIFA M-Check is developed which can effectively control various secondary bacterial infections effectively in single application during *Argulus* infection. For sustainable management of diseases, CIFA is also working on herbal based formulations for control of different diseases. In this line an effective herbal-based formulation (CIFA L-Check) that effectively controls argulosis is developed. CIFA L-Check kills all the life stages of the parasites very effectively. The reduced burden of parasite is immediately being reflected in terms of good skin luster and growth of fish, thereby increasing the economic return. Further, it has negligible effect on plankton biomass. The fish mortality is also reduced drastically with its application.

ICAR-CIFA is constantly working on the characterization of host immune elevation post immunostimulant incorporation. Several immunostimulants were studied for this purpose *i.e.* lactoferrin, beta-1,3 glucan, levamisole, vitamin C, etc. Immunoboost-C an immunostimulant is developed by the institute to improve brood fish health and seed production in carps. It modulates fish immunity against microbial diseases and has been proven through extensive trials conducted at many aquaculture regions in India. It is also given to spawn, fry and fingerlings through bath treatment during seed transport. Dietary supplementation of zinc oxide (ZnO) and selenium (Se) nanoparticles (NPs) stimulates immunity and enhances resistance to bacterial infection. In addition, developed "CIFA-Brood Vac" and tested its efficacy in female Amur carp broodfish. Fish exhibited 30% more spawn survivability in CIFA-Brood Vac administrated group, as compared to control group. Further, an array of antimicrobial peptides and their disease protection in carps have been looked into for therapeutic efficacy. Several diagnostics are also developed for the rapid diagnosis of different fish diseases. Developed RT-PCR based diagnostic to identify *M. rosenbergii* nodavirus and extra small virus associated with white tail disease with high degree of sensitivity and specificity. In addition, molecular based diagnostics were developed for koi herpes virus and spring viraemia of carp. Diagnostic kits developed for diagnosis of bacterial diseases of freshwater fishes include: i) Dot ELISA kit ii) Indirect ELISA kit, iii) Spot agglutination kit, iv) Antigen captured ELISA kit and v) Competitive ELISA kit for diagnosis of several pathogenic bacteria in freshwater fish species. A method for the identification and differentiation of two important argulus species *Argulus siamensis* and *Argulus japonicus* is also available. Further the laboratories have developed several ELISA-based systems for antioxidant molecules of fish using recombinant based protein antibodies, lectins from prawn etc.

Freshwater fish disease surveillance in Odisha and Andhra Pradesh

Under 'National Surveillance Programme on Aquatic Animal Diseases (NSPAAD)' the Central Institute of Freshwater Aquaculture (CIFA), Bhubaneswar was assigned to conduct active and passive surveillance for freshwater fish diseases in two states viz., Odisha and Andhra Pradesh in Phase I and for Odisha in Phase II. Under an active surveillance programme, this institute looked into the screening of transboundary viral pathogens viz., koi herpes virus, spring viraemia of carp virus and *Macrobrachium rosenbergii* nodavirus in the freshwater aquaculture sector in Khordha, Puri, Nayagarh, Balasore, Bhadrak, Jajpur, Kendrapara, Cuttack, Jagatsinghpur, Sambalpur districts of Odisha and West Godavari, East Godavari, Nellore, Guntur and Krishna districts of Andhra Pradesh. The passive surveillance has been generating information on important diseases listed in 'Diseases of National Concern" for both states. In April 2022 the second Phase of the project was sanctioned under Pradhan Mantri Matsya Sampada Yojana (PMMSY). In this new phase of the project, ICAR-CIFA has been given the responsibility of looking at active and passive disease surveillance of state Odisha. During the first phase of the project a comprehensive disease scenario of eastern India was generated through passive surveillance. Among different disease incidences, parasitic diseases were found to be the major contributor accounting for 81.93% of cases followed by bacterial diseases of 9.16%, mixed bacterial and parasitic diseases of 7.63% and viral diseases of 1.27%. Among different parasitic cases mixed parasitic infections (38.81%) were found to be the premier cause of disease, followed by argulosis (23.6%), dactylogyrosis (16.14%), myxosporean infections (10.86%), trichodinosis (3.41%), ichthyophthiriasis (2.17%) and other parasitic infections (4.96%). Several bacterial pathogens have been associated with causing disease and mortality in cultured fishes. Among bacterial infections, aeromonads group account for major disease causing agents. Besides, a few emerging bacterial pathogens like *Klebsiella pneumoniae* and *Proteus mirabilis* were identified to cause fish mortalities. Emerging viral pathogens like carp edema virus, cyprinid herpes virus-2 and infectious spleen and kidney necrosis virus were also identified from mortality cases of koi carp, goldfish and different ornamental fishes, respectively. Recently tilapia lake virus and tilapia parvovirus identified as etiological agent for causing large scale mortality in tilapia farms. Additionally, few parasites namely, *Metanophrys sinensis* from prawn, *Glossiphonia complanata* from pearl mussel and *Dactylogyrus scorpius* from rohu were identified and reported first time in the country.

Frontier research in aquaculture

Whole genome sequencing

The aquaculture field has experienced a revolution in genomic research owing to advances in sequencing technology. A genome encompasses all the DNA content, both coding, and non-coding, within an organism. The year 2011 marked a significant milestone when the first genome of Atlantic cod was successfully sequenced. Since then, the genomes of many significant aquaculture species have been sequenced and made available in public databases. In 2009, the Fish10K Genome Project was launched with the ambitious goal of sequencing and assembling genomes of approximately 10,000 vertebrate species, including 4,000 fish species (Bernardi *et al.*, 2012). Among the cyprinid species, seven important carp genomes that have been sequenced. Among them ICAR-CIFA have sequenced the whole genome of rohu carp (*Labeo rohita*) and catla (*Labeo catla*) using NGS technologies and reported the size of 1482 Mb and 1000 Mb, respectively (Sahoo *et al.*, 2020; Das *et al.*, 2020). This serves as an important resource for research on the comparative biology, evolution, and genomes of cyprinid species. Complete mitochondrial genome sequencing of various carp species such as *L. catla* of 16,597 bp in length (Kibria *et al.*, 2020), *L. rohita* of 16,606 bp in length (Das *et al.*, 2020), *C. mrigala* of 16,594 bp in length (Bej *et al.*, 2013) has been reported. Mitochondrial DNA (mtDNA) finds extensive applications in conservation genetics and phylogenetic analysis of numerous freshwater fish species. Integration of transcriptomics, epigenomics, and other functional genomics techniques with whole-genome sequencing (WGS), researchers have made a significant progress in identifying and characterizing genes linked to specific traits (Misra *et al.*, 2019).

Marker resources and their applications

Molecular markers are DNA sequences in the gene or genome that show genetic alterations at a particular location leading to genetic variation. This depicts the differentiation in the form of DNA polymorphism between organisms, species, population genetics studies at population levels that can be used for mapping and identifying individuals. In carp species, several genetic markers were identified for genetic characterization, assessment of genetic diversity, parentage assignments, species detection, population structure analysis, evolutionary study, linkage map construction, and selective breeding. It includes RAPD, RFLP, AFLP, microsatellites (SSRs), ESTs and SNPs, etc. Genetic markers also have been used for the conservation of genetic resources, stock differentiation, finding quantitative trait loci (QTL), and fisheries management. RAPD and AFLP have limited applicability due to their drawbacks, including dominant nature and irreproducible findings, respectively. On the other hand, genetic markers such as microsatellites, single nucleotide polymorphisms (SNPs), expressed sequence tags (ESTs), and mitochondrial (mtDNA) markers have become prevalent in aquaculture-related research. These new markers offer more reliable and diverse options for studying genetic variations in aquatic species. Now a days microsatellite and SNP markers have been widely used for population genetic structure analysis, QTL mapping, stock characterization, marker-assisted selection and genomic selection.

Genomic selection (GS) is an advanced molecular breeding method in aquaculture that uses markers, like single-nucleotide polymorphisms (SNPs), to enhance the accuracy of breeding values. This method harnesses genomic relationships and markers associated with essential genes, with its application steadily expanding across various aquaculture species. GS demonstrates efficacy in refining critical traits such as growth and disease resistance, driven by the overarching objective of bolstering sustainability and fortifying resilience against the impacts of climate change. Despite its considerable potential, the widespread integration of GS remains somewhat constrained, though the prospect of cost-effective strategies holds promise for broader implementation. Certain desirable traits (i.e., feed efficiency, disease resistance, fillet /carcass yields, and flesh quality), challenging to measure directly and require evaluations of siblings. In contrast to classical pedigree-based selection, GS excels at capturing within-family genetic variation, thereby enhancing the overall genetic response. GS also endeavours to reduce the generation interval and mitigate the challenges associated with inbreeding. The comprehensive GS program encompasses the establishment of breeding populations, individual phenotyping, estimation of genomic estimated breeding values (gEBV), trait-based selection and the development of training populations involving genotyping.

Genome editing

Genome editing also known as gene editing is a method that involves making specific changes to the DNA or region of genome of a cell or organism. The genome editing technologies like transcriptional activator-like effector nucleases (TALENs) and zinc finger nuclease (ZFN) were early methods for targeted genome editing. However, they have been largely replaced by the more efficient and versatile CRISPR/Cas9 system *i.e.* CRISPR (clustered regulatory interspaced short palindromic repeats)/CRISPR associated nucleases (Cong *et al.*, 2013). CRISPR/Cas9 allows for precise editing of genomes, including gene knockout, activation, inhibition, and even epigenetic modifications, revolutionizing biological research, and offering new possibilities for understanding and modifying DNA. Selective breeding in cultured species is constrained by factors like low heritability of traits, long generation-intervals, the complexity of targeting multiple traits, and limited genetic variation in broodstock. CRISPR/Cas9 offers the potential for rapid genetic improvement in aquaculture by targeting specific traits like sterility, growth, and disease resistance. CRISPR/Cas9 and TALEN genome editing techniques have been effectively used in several aquaculture species, including cyprinids such as common carp, rohu, and grass carp. The availability of reference genomes will allow better design of guide RNAs for the CRISPR-Cas9 vector, which results into minimized off-target effects. Thus the CRISPR-Cas9 techniques hold great potential for targeting complex traits like immunity and disease resistance, crucial for aquaculture improvements and disease management.

ICAR-CIFA is frontrunner in the field of genome editing of fish. The single guide RNAs targeting perforin 1a, perforin 1b and mucin exon were designed for zebrafish and *L. rohita* and co-injecting with cas9 mRNA at one cell stage of zebrafish and *L. rohita* embryos for producing gene knockouts. The heteroduplex assay and high resolution melt curve analysis confirm the gene knockout for perforin 1a and perforin 1b in zebrafish and *L. rohita* and F_0 generation were produced. The both perforin 1a and perforin 1b F_0 adult zebrafish were screened for positive mutation using heteroduplex assay and HRM analysis. The perforin 1a F_0 zebrafish adults resulted positive for mutation in all three zebrafish perforin isoforms (zP1.2, zP1.7 & zP1.8) were selected for producing F_1 generation and the perforin 1b F_0 zebrafish adults resulted positive for mutation in zebrafish perforin isoform (zP1.5) selected for producing F_1 generation. These screened F_0 zebrafish adults were crossed with wild type for developing perforin 1a and perforin 1b F_1 zebrafish generation. The F_1 embryos were raised to adult and fin clips were collected and DNA isolated. All the F_1 adults were screened for gene mutation. Interestingly the perforin 1a F_1 zebrafish were shown different pattern for gene mutation among each other as same observed in F_0. The F_1 fish having all three perforin mutations were selected for developing F_2 generation. The perforin 1b F_1 adults resulted positive for mutation were selected for producing F_2 generation.

AMR and One health

Antimicrobial resistance (AMR) is the ability of microorganisms to resist the effects of antimicrobials. As a result, standard treatments become ineffective, infections persist and may spread to others. Under INFAAR-Indian Network for Fisheries and Animal Antimicrobial Resistance, ICAR-Network Programme on Antimicrobial Resistance (AMR) in Aquaculture & Fisheries was launched on 24th March, 2018 at ICAR-NBFGR, Lucknow. Among 8 fisheries Institutes under ICAR, ICAR-CIFA, Bhubaneswar is working on the objectives to determine the extent of antimicrobial resistance (AMR) in selected bacterial species of public health significance isolated from fish samples collected from aquaculture ponds from selected regions in Odisha and Andhra Pradesh. At ICAR-CIFA during 2018-23, survey and sampling was conducted in 708 freshwater fish farms in Odisha and Andhra Pradesh. During the survey, both fish and water samples were collected and subsequently processed for the isolation of *Aeromonas* sp., *E. coli*, and *Staphylococcus* sp. A total of 2992 bacterial isolates were isolated and processed for antimicrobial susceptibility testing comprising 938 isolates of *Aeromonas* sp., 1021 isolates of *E. coli*, and 1033 isolates of *Staphylococcus* sp. *Aeromonas* sp. isolates showed the highest resistance to cefoxitin (49.1%) followed by cefotaxime (25.8%), trimethoprim/sulfamethoxazole (21.7%), ceftazidime (11.7%), etc. *E. coli* isolates were resistant to ampicillin (19.3%), followed by nalidixic acid (17.9%), cefoxitin (17.1%), tetracycline (16.7%), etc. *Staphylococcus* sp. isolates showed the highest resistance to penicillin (78.3%), followed by cefoxitin (34.9%), tetracycline (11.3%), trimethoprim/sulfamethoxazole (10.3%), etc.

Alternatives to antibiotics in aquaculture

Antibiotic resistance has forced the researchers to look into other alternatives for management and control of diseases in aquaculture sector. Putative lactic acid bacteria (LAB) diversity in freshwater fish has been studied for which 76 strains of LAB were isolated from intestines and identified by phenotypic tests and 16S rDNA gene sequencing. Molecular identification revealed *Lactobacillus plantarum, Lactobacillus pentosus, Lactobacillus fermentum, Lactobacillus delbrueckii* sub sp. *bulgaricus, Lactobacillus brevis, Lactobacillus reuteri, Lactobacillus salivarius, Pediococcus pentosaceus, Pediococcus acidilactici, Weissella paramesenteroides, Weissella cibaria, Enterococcus faecium* and *Enterococcus durans. Lactobacillus plantarum* was found to be the dominating strain in hindgut. Selective putative strains with good probiotic attributes including acid and bile tolerance capability, ability to produce bacteriocins and suppress pathogen growth under *in vitro* conditions, could be successfully used for sustainable and environment friendly aquaculture (Maji *et al.*, 2016). Lactic acid bacteria are fastidious in nature and also present in less quantity in freshwater environment. Hence, they require specific isolation procedures. A study reported double enrichment cum selective isolation-based protocol for successful isolation of lactic acid bacteria (LAB) from freshwater fish intestine. Dietary implications of probiotics have been successful in imparting monoprophylaxis and disease resistance in fish. A dietary consortium of putative Lactic acid bacterial multistrain probionts proved to be a potent growth promoter and immune stimulator in rohu and consequently conferring better protection against *A. hydrophila* by regulating the expression of immune regulatory genes (Maji *et al.*, 2017). In addition to probiotic formulations several herbal based formulations are also under trial as alternative to antibiotics. In addition to this to reduce the occurrence of diseases and to manage diseases ICAR-CIFA is actively working on development of different vaccines.

Digital Outreach

MatsyaSetu – Mobile App

ICAR-CIFA with funding support from National Fisheries Development Board (NFDB), Hyderabad has recently developed and launched a mobile android app called “MatsyaSetu”. MatsyaSetu app has species-wise/ subject-wise self-learning online course modules, where renowned aquaculture experts explain the basic concepts and practical demonstrations on breeding, seed production and grow-out culture of commercially important fishes like carp, catfish, scampi, murrel, ornamental fish, and pearl farming. Better management practices to be followed in maintaining the soil and water quality, feeding and health management in aquaculture operations have also been provided on the course platform. So far, more than 40,000 users have downloaded this app and are using the technology. ICAR-CIFA has also released an upgraded version of MatsyaSetu 2.0 with many additional features including Aqua Bazar, farmers/buyers/seller's registration, farmer's advisories etc.

TreatMyFish – Mobile App

Fish health management and diagnosis is a key component of sustainable freshwater aquaculture production. Many of the states across the country don't have diagnostic laboratory facilities and also lack experts to render services at the farmers' point. ICT tools pose a great potential to provide need-based specific advisory services to fish health problems of aquafarmers. In this background, ICAR-CIFA has come out with an Android Mobile app named "TreatMyFish" to foster the reach of timely fish health management information and provide interactive real-time solutions for aquafarmers across the country. TreatMyFish is a mobile app providing information on fish health management, including disease diagnosis and control measures. These applications enhance learning, market connectivity, and fish health management in aquaculture.

Conclusion

ICAR-CIFA with the support of Department of Fisheries, Government of India and other state departments is tirelessly working on developing farmer friendly technologies and supporting the mission of GoI for brining sustainable aquaculture development by expanding fish production through innovative programs and focusing on production management, species and system diversification, genetic improvements, and technology adoption. Our initiatives include broodstock enhancement, year-round seed production, cost effective feed development, disease control and therapeutic applications, and impact assessment. Capacity building is a priority, involving scientists, farmers, and stakeholders to ensure global competitiveness and food security. By 2030, Govt. of India aims to produce 25 million tonnes of fish annually, with 80% from aquaculture. Our strategy involves development of genetically improved priority species through cost-effective methods. Sustainable aquaculture practices through policy support and effective technology transfer and committed to optimizing resources, enhancing market links, and improving post-harvest processing. With these strategies, ICAR-CIFA is poised for brining success in blue revolution mission in the country.

Further Reading

Ayyappan, S., Moza U., Gopalakrishnan, A., Meenakumari B., Jena, J.K., Pandey, A.K.,2011. Handbook of fisheries and aquaculture. DKMA, ICAR, New Delhi

Bej, D., Sahoo, L., Das, S. P., Swain, S., Jayasankar, P., and Das, P. (2013). Complete mitochondrial genome sequence of Cirrhinus mrigala (Hamilton, 1822). Mitochondrial DNA 24, 91–93. doi:10.3109/19401736.2012.722998

Bernardi, G., Wiley, E. O., Mansour, H., Miller, M. R., Orti, G., Haussler, D., *et al.* (2012). The fishes of Genome 10K. Mar. Genomics 7, 3–6. doi:10.1016/j.margen.2012.02.002

Cong, L., Ran, F. A., Cox, D., Lin, S., Barretto, R., Habib, N., *et al.* (2013). Multiplex genome engineering using CRISPR/Cas systems. J Sci. 339, 819–823. doi:10.1126/science.1231143

Das, P., Sahoo, L., Das, S. P., Bit, A., Joshi, C. G., Kushwaha, B., *et al.* (2020). De novo assembly and genome-wide SNP discovery in rohu carp, *Labeo rohita.* Front. Genet. 11, 386. doi:10.3389/fgene.2020.00386

Dash, P., Sahoo, P. K., Gupta, P. K., Garg, L. C., & Dixit, A. (2014). Immune responses and protective efficacy of recombinant outer membrane protein R (rOmpR)-based vaccine of *Aeromonas hydrophila* with a modified adjuvant formulation in rohu (*Labeo rohita*). Fish & shellfish immunology, 39(2), 512-523.

FAO. 2022. The State of World Fisheries and Aquaculture 2022. Towards Blue Transformation. Rome, FAO. https://doi.org/10.4060/cc0461en

Gjerde, B., Reddy, P. V., Mahapatra, K. D., Saha, J. N., Jana, R. K., Meher, P. K., ... & Rye, M. (2002). Growth and survival in two complete diallele crosses with five stocks of Rohu carp (*Labeo rohita*). Aquaculture, 209(1-4), 103-115.

https://cifa.nic.in/

https://pmmsy.dof.gov.in/

Jayasankar, P., 2018. Present status of freshwater aquaculture in India-A review. Indian Journal of Fisheries, 65(4), pp.157-165.

Kar, B., Mohapatra, A., Mohanty, J., & Sahoo, P. K. (2017). Evaluation of ribosomal P0 peptide as a vaccine candidate against *Argulus siamensis* in *Labeo rohita.* Open Life Sciences, 12(1), 99-108.

Mahapatra K. D., Jayasankar P, Saha JN, Murmu K, Rasal AR, Nandanpawar P, Patnaik M, Sundaray JK, Sahoo PK (2016). Jayanti Rohu: Glimpses from the Journey of First Genetically Improved Fish in India, 1st edn. ICAR-CIFA, Bhubaneswar

Mahapatra KD, Gjerde B, Sahoo PK, Saha JN, Barat A, Sahoo M. *et al.* (2008) Genetic variations in survival of rohu carp (*Labeo rohita*, Hamilton) after *Aeromonas hydrophila* infection inchallenge tests. Aquaculture 279: 29–34.

Mahapatra, K.D., Sahoo, L., Saha, J. N., Murmu, K., Rasal, A., Nandanpawar, Das, P. and Patnaik, M. (2018). Establishment of base population for selective breeding of catla (*Catla catla*) depending on phenotypic and microsatellite marker information. Journal of genetics, 97, 1327-1337.

Misra, B. B., Langefeld, C., Olivier, M., and Cox, L. A. (2019). Integrated omics: tools, advances and future approaches. J. Mol. Endocrinol. 62, R21–R45. doi:10.1530/jme-18-0055

MoFAHD, 2023. Annual Report 2022-2023, Department of Fisheries, Ministry of Fisheries, Animal Husbandry & Dairying, Govt. of India, New Delhi

Rasal, A., Patnaik, M., Murmu, K., Sundaray, J. K., Vasam, M., Swain, J. K., & Mahapatra, K. D. (2022). A comparative analysis of Passive Integrated Transponder (PIT) tagging in selective breeding programme of improved rohu (Jayanti) and catla. Aquaculture Reports, 26, 101284.

Sahoo, L., Das, P., Sahoo, B., Das, G., Meher, P. K., Udit, U. K., *et al.* (2020). The draft genome of Labeo catla. BMC Res. Notes 13, 411. doi:10.1186/s13104-020-05240-w

Sahoo, P. K., Rauta, P. R., Mohanty, B. R., Mahapatra, K. D., Saha, J. N., Rye, M., & Eknath, A. E. (2011). Selection for improved resistance to *Aeromonas hydrophila* in Indian major carp *Labeo rohita*: Survival and innate immune responses in first generation of resistant and susceptible lines. Fish & Shellfish Immunology, 31(3), 432-438.

Saikia, D., Bhuyan, M. K., & Das, N. (2020). Growth performance of Jayanti Rohu and Amur Common carp in extensive polyculture system. Journal of Krishi Vigyan, 9(si), 152-155.

Swain S., Ferosekhan S. 2022. Present Status and Future Scope of Freshwater Aquaculture Sector in India. In: Souvenir of IFO-2022 (Eds. Das *et al.*)ICAR-CIFRI, Barrackpore, ISBN: 8185482446

Yadav, S. K., Dash, P., Sahoo, P. K., Garg, L. C., & Dixit, A. (2018). Modulation of immune response and protective efficacy of recombinant outer-membrane protein F (rOmpF) of *Aeromonas hydrophila* in *Labeo rohita*. Fish & Shellfish Immunology, 80, 563-572.

Yadav, S. K., Dash, P., Sahoo, P. K., Garg, L. C., & Dixit, A. (2021). Recombinant outer membrane protein OmpC induces protective immunity against *Aeromonas hydrophila* infection in *Labeo rohita*. Microbial Pathogenesis, 150, 104727.

3

Pioneering Initiatives in Marine Fisheries Advancing Climate Research for Sustainable Ecosystems and Aquaculture

Grinson George* and A. Gopalakrishnan

ICAR-Central Marine Fisheries Research Institute, Kochi – 682 018. Kerala

**Email: director.cmfri@icar.gov.in*

Introduction

Climate change has emerged as a paramount global concern, representing threat to food and nutritional security, particularly for the growing global population. While the impacts of climate change are felt worldwide, certain regions, including India, stand out as more vulnerable. India's susceptibility to the implications of climate change is highlighted by its large and dependent population relying significantly on fisheries. The obscure interaction between climate change and fisheries poses a complex challenge, encompassing ecological, economic, and social dimensions. As the climate continues to evolve, the repercussions on marine ecosystems and fisheries could have far-reaching consequences, underscoring the urgent need for comprehensive research and adaptive strategies to ensure the resilience of these critical sectors. CMFRI demonstrates a comprehensive commitment to advancing knowledge and practices in marine ecosystems. From meticulously studying the phenology, distribution, and trophodynamics of climate-vulnerable teleost fishes to employing sophisticated climate change modeling techniques. The integration of field data into a geospatial database enhances the understanding of coastal ecosystems. In a multifaceted approach, CMFRI combines field and laboratory studies, utilizing technologies to monitor habitat usage and species diversity, fostering both scientific understanding and public engagement. The institute's emphasis on vulnerability assessments, hazard atlases, and Climate Change Risk Index calculations reflects understanding of the intricate dimensions of climate-related impacts. CMFRI's commitment extends to sustainable practices, exemplified through climate-resilient farming initiatives such as Integrated Multi-trophic Aquaculture (IMTA), seaweed culture, and the strategic establishment of artificial reefs. CMFRI not only comprehends the challenges posed by climate change but actively contributes to innovative and sustainable solutions in fisheries and aquaculture, emphasizing its dedication to the well-being of marine ecosystems.

Phenology, Distribution and Trophodynamics

The research conducted at the Central Marine Fisheries Research Institute (CMFRI) in India, focusing on the phenology, distribution, and trophodynamics of climate-vulnerable teleost fishes, represents a comprehensive exploration of critical aspects of marine ecology. The study encompasses trophodynamics, phenology, and distribution, exploring the complex ecological interactions, seasonal patterns, and geographical distribution of teleost fishes susceptible to climate variations. A noteworthy dimension of this research involves the identification of locally adapted sardine stocks in the vast expanse of the Indian Ocean, utilizing the ddRAD sequencing techniques. This molecular approach provides a detailed and fine-grained understanding of the genetic variations within sardine populations, contributing to the development of targeted conservation and management strategies. Expanding the scope, the research further encompasses trophodynamics, phenology, and reproductive parameters of 29 climate-vulnerable teleost fishes from diverse regions.

This broad approach facilitates a comprehensive examination of the ecological and biological dynamics of these species, shedding light on their life cycles, reproductive behaviours, and responses to environmental changes. Understanding these aspects is crucial for predicting and mitigating the impacts of climate change on marine biodiversity. The distribution patterns of five marine species in the Arabian Sea and Bay of Bengal have been studied over a specific period, and the resulting geocoded data has been effectively mapped in a GIS platform. This spatial analysis provides valuable insights into the habitats and movement patterns of these species, aiding in the formulation of conservation and management strategies. By employing geospatial technology, CMFRI enhances its capacity to visualize, analyze, and interpret complex ecological data, contributing to a more informed and effective conservation approach. Moreover, the research extends its focus to the climatic resilience and adaptive potential of Clupeid fishes across the world oceans, employing a comparative mitogenomics approach. This molecular investigation into the mitochondrial genomes of Clupeid fishes provides a deeper understanding of the genetic mechanisms underlying their adaptability to changing climatic conditions (Fig.1). CMFRI's research on the phenology, distribution, and trophodynamics of climate-vulnerable teleost fishes represents a holistic and interdisciplinary approach to marine ecology. From advanced molecular techniques like ddRAD sequencing and comparative mitogenomics to the application of geospatial analysis, the institute employs a diverse array of methodologies to unravel the complexities of marine ecosystems.

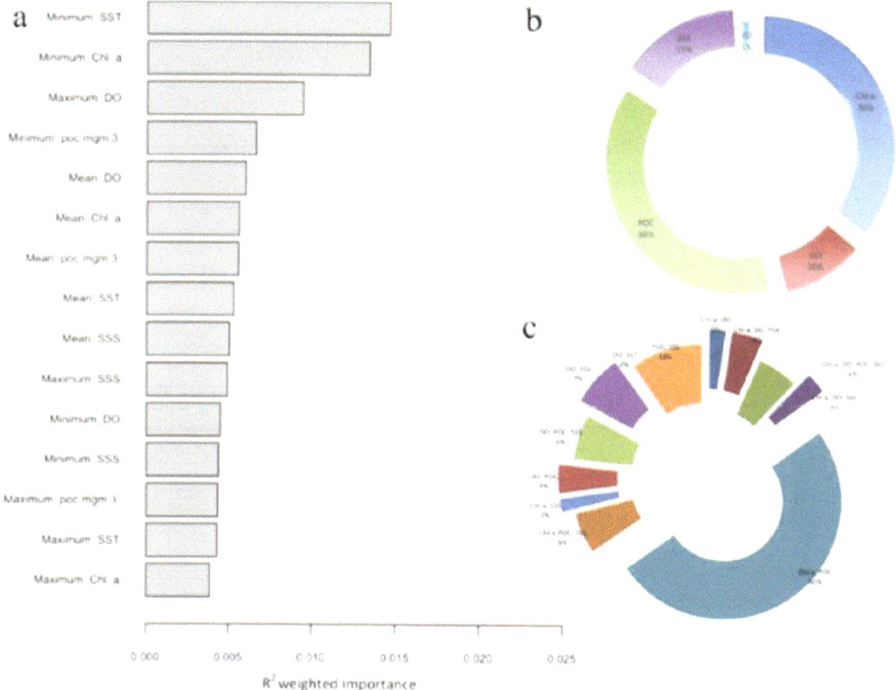

Fig. 3.1: Contribution of each environmental variable in explaining genetic variation across *S. longiceps* populations generated from gradient forest analysis

Carbon footprint and blue carbon assessment

The comprehensive assessment of the carbon footprint of Indian marine fisheries conducted by ICAR-CMFRI involved an in-depth analysis of greenhouse gas equivalent (GHG) emissions from diverse fishing activities, encompassing construction and fabrication. The developed methodology, standardized by ICAR-CMFRI and implemented from 2013 to 2018, offers a unique approach to evaluating multi-sector, multi-gear marine fishing activities in India, with a focus on pre-harvest, harvest, and post-harvest scenarios for various fishing methods. This methodology integrates standard processes with indigenous modifications, serving as a model adaptable for countries with diverse small-scale mechanized/motorized fishing sectors (Fig.2). A global study reported an average release of 2.2 kg of CO_2 equivalent emissions (CO_2e) to catch

1 kg of fish. However, in India's mechanized fishery sector, the average emission was notably lower at 1.47 kg CO_2e per kg of fish, representing a 33% reduction compared to the global estimate. Similarly, for fishing operations in India encompassing mechanized, motorized, and indigenous sectors, the average emission stood at 1.52 kg CO_2e per kg of fish, indicating a 30% reduction compared to the global estimate. Over the past decade, GHG emissions from Indian trawlers have consistently remained below the global average, initially at 16.3% less and now at 17.7% less than the global average, despite increased emissions globally.The pivotal role of mangroves in climate change mitigation arises from their substantial contribution to capturing and storing carbon, emphasizing their significance as vital blue carbon ecosystems. This study delves into assessing the monetary value of services provided by India's mangrove ecosystems, specifically focusing on their contribution to reducing CO_2 emissions through carbon sequestration. Utilizing a comprehensive methodology and field-level data from Kerala and Andhra Pradesh, combined with secondary data from coastal states and Union Territories across India, the study evaluated carbon stocks in above-ground, below-ground (root), and sediment components. Converting estimated total carbon stocks into CO_2equivalence and applying standard estimates, the study determined their corresponding monetary value. Findings revealed a combined extrapolated total carbon stock of 59,463,228 tonnes across India's coastal regions, translating to a total CO_2 equivalence of 217 million tonnes. Notably, the cumulative monetary value of carbon sequestration efforts within Indian mangrove ecosystems amounted to an impressive 18,716 million US dollars. Furthermore, a model mangrove farm in Tuticorin, employing an allometric approach, estimated carbon stocks in above-ground and below-ground biomass, unveiling variations in carbon stock over time. Additionally, seaweed farming in Tuticorin explored carbon sequestration potential, with GIS data and satellite imagery estimating seaweed farming areas and biomass production. The assessment of net carbon sequestration potential revealed fluctuations over different years, providing insights into factors influencing seaweed production and its role in carbon sequestration. Collectively, these studies significantly contribute to the understanding of carbon dynamics in coastal ecosystems and the potential for mitigating climate change through blue carbon initiatives.

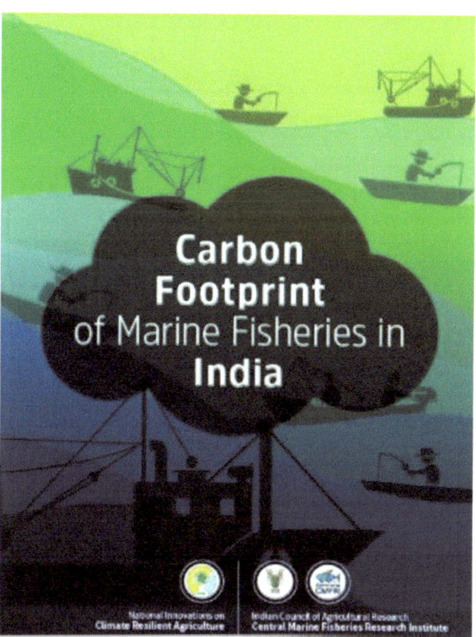

Fig. 3.2: Carbon footprint study of marine fisheries in India

Geospatial Assessment for Coastal Ecosystems

In a determined effort to integrate field data into a comprehensive geospatial database, CMFRI collaborated with the Indian Space Research Organisation (ISRO) and the National Innovations in Climate Resilient Agriculture (NICRA) program, employing the ISRO-CMFRI-NICRA wetland mobile application. This initiative involved systematic and continuous samplings across 96 wetlands situated around coastal villages in the states of Kerala, Karnataka, and Tamil Nadu. The data collected through this extensive fieldwork became a vital component of the geospatial database, providing valuable insights into the dynamic nature of coastal ecosystems. A significant aspect of the geospatial assessment focused on evaluating temporal changes in seagrass coverage within Kalpeni lagoon. This analysis spanned the years from 2003 to 2020 and was conducted using Landsat data on the Google Earth Engine platform (Fig 3). The objective was to recognise alterations in seagrass coverage over time and to understand the influence of selected climate factors on this vital coastal ecosystem. The findings of this study revealed a decline of over 99% in seagrass coverage, deteriorating health of the seagrass ecosystem in the region. The continuous samplings in the wetlands allowed for the creation of a detailed geospatial map, enriching the understanding of the spatial distribution of various ecological features.

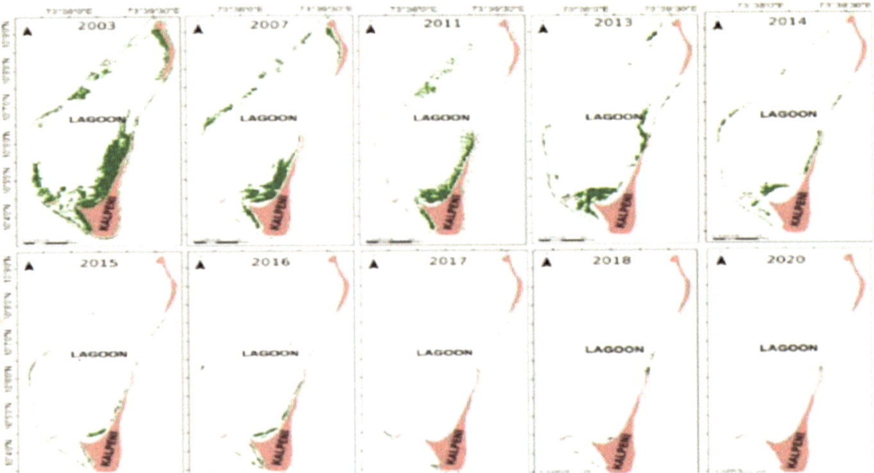

Fig. 3.3: The change in the seagrass coverage along the years from 2003 to 2020

The effective management of fisheries can greatly benefit from the utilization of geographic information system (GIS) tools. These tools facilitate habitat mapping, georeferencing fish catch and fishing effort data, and linking catch information to oceanographic and biochemical parameters. Historically, the range of Indian fishermen was limited due to the constraints in the mechanization of fishing vessels. Catch reports were typically associated with the nearby coastal area of a designated landing center. However, with the widespread adoption of powerful engines, fishing evolved into a more professional pursuit, with mechanized boats undertaking extended journeys lasting weeks.The shift towards extensive fishing activities resulted in an immediate challenge, the lack of accurate information about the specific locations where catches were obtained. Precise identification of fishing grounds became possible with the use of GPS devices, but retroactively assigning geographical coordinates to historical data lacking latitude and longitude details remained a challenge. Typically, landing records include information about bearing and the distance covered by the surveyed craft. Using the Haversine formula to determine destination coordinates based on distance and bearing from the starting point coordinates, probable latitude and longitude of fishing grounds can be estimated. Passive geo-referenced species distribution data was plotted in a GIS platform and overlaid with India's coastal line spatial dataset. Spatial analysis tools refined the datasets and eliminated false information. Once the datasets were prepared, they were superimposed with various environmental variables derived through remote sensing, such as ocean temperature, salinity, dissolved oxygen, and chlorophyll. A well-designed, easily modifiable database was created. A modeling attempt was made to establish a relationship between the dichotomous dependent variable (presence/absence) of the species and environmental drivers. The Ensemble model with a logistic link function was employed for this purpose. The results indicated a good fit with a significant influence of climatic factors on species presence/absence. The fitted models were then used to project the probable distribution of the species along the west coast.

Field and Laboratory Studies on Species Adaptation to Climate Change Resilience

Broad and innovative study was conducted, integrating both field and laboratory investigations to assess the mitigation potential of mangroves and coastal wetlands within India's diverse coastal ecosystems. The research employed non-invasive Remote Underwater Video (RUV) technology, which played a pivotal role in monitoring habitat usage, species abundance, and diversity. This approach not only contributed to the scientific understanding of coastal ecosystems but also engaged the public in the conservation discourse. By laying the groundwork for evaluating conservation measures, the study aimed to enhance the resilience of these crucial habitats. Laboratory experiments were undertaken to assess the thermal tolerance of the silver pompano fish species. The findings revealed that acclimation to higher temperatures significantly increased their thermal tolerance, positioning these resilient fish as potential candidates for integration into climate-smart aquaculture systems. The combined insights from field and lab data provided a holistic perspective on coastal ecosystem conservation and species adaptation to climate change, offering valuable information for sustainable management practices. Multifaceted study investigated the complexities of climate-driven changes in small pelagic fish populations, employing a meticulous approach to understand and adapt to environmental shifts.

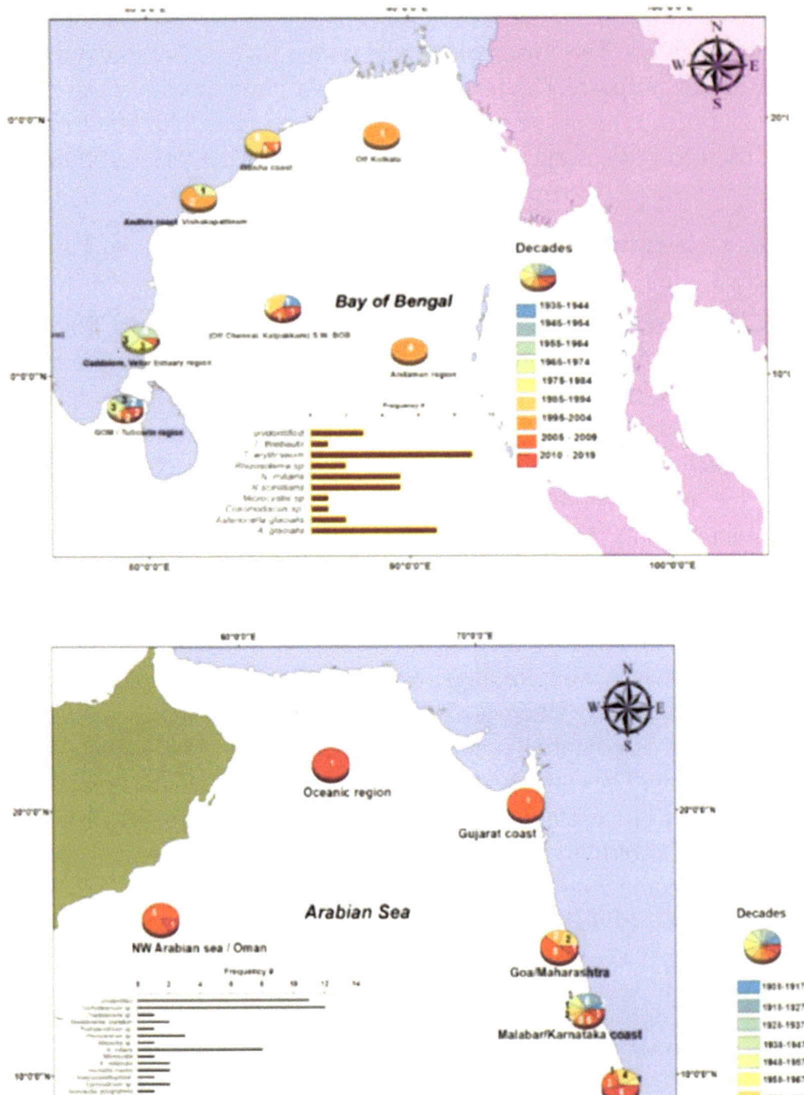

Fig. 3.4: Secondary data of bloom events across Arabian Sea (AS) (1908 to 2015) and Bay of Bengal (BoB) (1935-2019) was catalogued to identify spatio-temporal variability

Notably, the research identified the optimal acclimatization temperature (Topt) for the *Trachinotus mookalee* as 26°C, showcasing a thermal tolerance range between 12.66°C and 43.22°C. Additionally, the study observed a remarkable 100% survival rate for spiny lobsters (*Panulirus homarus)* exposed to gradual salinity changes between 17‰ and 44‰, as well as abrupt salinity exposures ranging from 26‰ to 41‰. The impact of environmental variables on resource abundance, potential yields, and spawning of selected marine fishes and crustaceans was systematically assessed. This comprehensive analysis contributed to the understanding of how these species respond to climate-related changes, providing crucial insights for fisheries management and conservation efforts. Harmful Algal Blooms (HABs) pose a significant risk to India's coastal resources, with repeated emergence of toxin producing HABs causing substantial losses in aquaculture as algal toxins accumulate in fishes beyond permissible levels. Non-toxic algae blooms can also cause harm through various means, including the substantial biomass achieved by certain blooms. As these blooms decompose, oxygen depletion occurs, resulting in widespread mortality of plants and animals in the affected area. Prolonged blooms of non-toxic algal species, even in the absence of toxins, can have detrimental effects by reducing light penetration to the bottom, adversely affecting submerged aquatic vegetation, and impacting coastal ecosystems. These vegetation areas, serving as critical nurseries for the offspring of commercially important fish and shellfish, further contribute to the impact on fisheries.

As part of a comprehensive study, secondary data of bloom events across the Arabian Sea (AS) from 1908 to 2015 and the Bay of Bengal (BoB) from 1935 to 2019 was catalogued to identify spatio-temporal variability (Fig.4.) In the AS, approximately a three-fold increase in HAB events is reported during the last two decades (31 HAB events) compared with the first two decades (10 HAB events). Similarly, in the BoB, there is an approximately two-fold increase in HAB events during the last two decades (14 HAB events) compared to the first two decades (6 HAB events).

This comprehensive analysis contributes valuable insights into the dynamics of HABs in the Arabian Sea and the Bay of Bengal, aiding in the development of effective strategies for coastal resource management. Furthermore, investigated the patterns and impacts of harmful algal blooms, specifically the green tide caused by *Noctiluca scintillans*, along the Gulf of Mannar. By separating the dynamics of these algal blooms, the research addressed the ecological consequences and potential mitigation strategies for this phenomenon, fostering a more resilient marine ecosystem.

Vulnerability Assessment for Climate Change Risk

Widespread vulnerability assessment was undertaken, aiming to identify and finalize various variables within Hazards, Exposure, and Socio-Ecosystem vulnerability dimensions for the calculation of the Climate Change Risk Index. This meticulous process involved a multifaceted examination of factors contributing to the vulnerability of coastal regions, providing understanding of the complex interplay between hazards, exposure, and socio-ecosystem dynamics. The outcome of this assessment would form the basis for informed decision-making and adaptive strategies in the face of climate change.

To visualize and communicate the vulnerability of coastal districts, Hazard Atlas was developed based on the hazard vulnerability index calculated for each identified risk. Five significant hazards with profound physical vulnerability implications were carefully chosen, including Cyclone Proneness, Flood Proneness, Shoreline Change, Sea Level Rise, and Heatwave. Each hazard underwent a detailed vulnerability index calculation, allowing for the creation of district and state-level atlases that vividly illustrated the degree of vulnerability to each specific hazard. This approach not only facilitated a targeted response to individual threats but also provided a complete view of the overall vulnerability landscape.The development of a Multi-hazards Index (MHI) emerged as a key component of the vulnerability assessment, offering a composite index showcasing the relative proneness of coastal districts to a multiplicity of physical hazards.

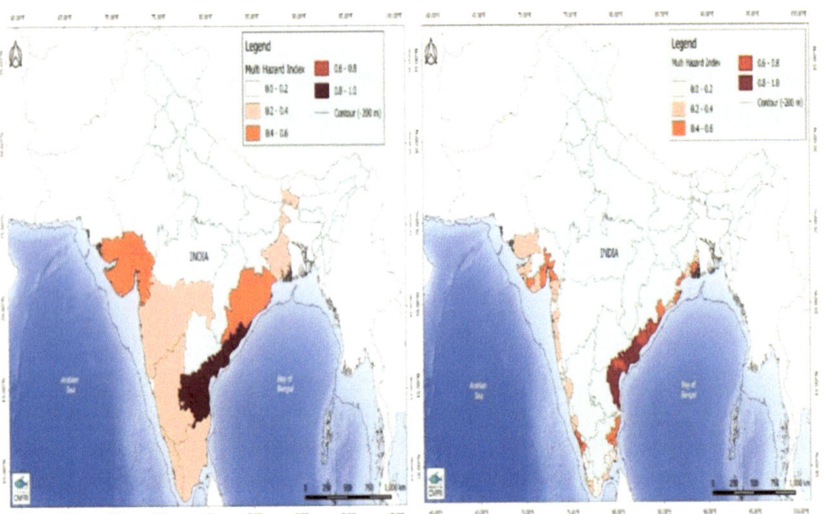

Fig. 3.5: Maritime state-wise and coastal district-wise multi hazards index map

The coastal state of Andhra Pradesh emerged with the highest score for the Multi-hazards Index (MHI), signifying the state's elevated susceptibility to physical hazards induced by climate change (Fig.5). Conversely, Goa exhibited the least proneness to these climate change-induced physical hazards, as evidenced by its lower index value. This comparative analysis provides valuable insights into the varying degrees of vulnerability across different regions. Further refining the assessment, the relative positions of coastal districts within the hazard atlas, based on the Multi-hazard Index, revealed vulnerability patterns. Sindhudurg district in Maharashtra state obtained the lowest score, indicating a relatively lower susceptibility to the identified hazards. Conversely, the Krishna district in Andhra Pradesh exhibited the highest score, signaling heightened vulnerability in this region. These localized vulnerability assessments contribute significantly to the development of targeted adaptation and resilience strategies, allowing for more effective risk management and climate change mitigation at the regional level.

Climate Resilient Farming Practices

CMFRI has been at the forefront of pioneering climate-resilient farming practices, introducing transformative technologies and methodologies that not only enhance productivity but also promote sustainability in the aquaculture sector.One of the landmark achievements has been the development and popularization of cage farming technology in India. With technical support from CMFRI, more than 3900 cages have been established, resulting in an impressive production of over 12,000 tonnes of harvested fishes. This initiative has not only significantly increased production efficiency but has also made aquaculture economically viable, with the cost of production reduced from Rs 350/kg to a more affordable Rs 210/kg. At the Kanyakumari coast, CMFRI initiated climate-smart farming practices, collaborating with Self-Help Groups (SHGs) in the region. Utilizing Geographical Indication (GI) square floating cages measuring 6 m x 6 m with four partitions, the initiative demonstrated remarkable success. Within 40 days, a fourfold growth was observed in seaweeds, showcasing the adaptability and efficiency of this innovative farming approach. The positive outcomes underline the potential for sustainable and lucrative seaweed cultivation in the context of climate resilience. CMFRI further advanced climate-resilient farming through the integration of sea cage culture of seabass with seaweed *Kappaphycus alvarezii* using the net tube method. After 10 months, this integrated approach yielded a seabass harvest of approximately 1144 kg, generating a total revenue of Rs. 4,91,420. This successful integration not only maximizes yield but also promotes ecological sustainability by harnessing the complementary relationships between different species within the aquatic ecosystem.The adoption of Integrated Multi-trophic Aquaculture (IMTA) represents another milestone in climate-resilient farming. CMFRI standardized and developed a package of practices for IMTA, combining 16 seaweed bamboo rafts with one cage.

Fig. 3.6: IMTA

The results demonstrated a remarkable 56% additional yield and an 18% increase in income for farmers. IMTA not only significantly improved water quality in cage culture sites when integrated with seaweeds but also reduced the chances of eutrophication. Furthermore, the carbon sequestration potential of cultivated seaweed in IMTA systems surpassed that of non-integrated rafts, underlining the environmental benefits of this innovative approach. In the field of seaweed culture, CMFRI has developed sea farming techniques for five species of seaweeds. Through GIS-based models, 317 potential seaweed farming sites, covering 23,950 hectares, have been identified. Monoline/ longline farming of *K. alvarezii* has proven to be successful, with wet seaweed production ranging from 6 to 10 tonnes per plot per crop (35 days). The dry seaweed yield, coupled with a market price of Rs. 40 per kg, translates to a substantial net income ranging from Rs. 12,000 to 16,000 per plot.

Artificial Reefs in India

The inception of organized efforts on Artificial Reefs (AR) in India dates to 1980s, marking a pivotal moment in marine conservation and ecosystem management. Since then, the initiative has evolved into a remarkable conservation strategy, contributing significantly to the enhancement of marine biodiversity and the promotion of sustainable fisheries. As of the present day, India boasts over 280 functional AR sites, strategically positioned along its vast coastal stretches. These sites collectively encompass an impressive 0.37 million square meters of AR area in Indian coastal waters. This expansive network of artificial reefs serves as a testament to the enduring commitment of marine scientists and conservationists to augmenting the natural habitat for marine life.

Fig. 3.7: The latest AR modules deployed by CMFRI

The success of India's AR program can be attributed not only to its scale but also to the meticulous protocols established for every stage of the AR lifecycle. Evolving from decades of research and hands-on experience, these protocols encompass crucial aspects such as site selection, design, fabrication, deployment, and impact assessment. Site selection is a meticulous process, involving the identification of locations that align with specific conservation objectives. These objectives include boosting fish populations, restoring degraded habitats, protecting sensitive marine ecosystems. The careful consideration of ecological factors ensures that AR sites become thriving havens for marine life. Designing artificial reefs demands a deep understanding of local marine ecology and the specific needs of targeted species. The structures are crafted to mimic natural substrates, providing attachment points for marine organisms to colonize. The fabrication of AR structures involves the use of durable, environmentally friendly materials that withstand the harsh marine environment. The deployment phase is a critical aspect of the AR program, requiring precision in placing structures to optimize their impact on marine life. Once deployed, these artificial reefs become a focal point for colonization by various species, fostering biodiversity and creating new opportunities for fisheries. Impact assessment is an ongoing process that evaluates the effectiveness of ARs in achieving their intended goals. This involves monitoring changes in marine biodiversity, fish populations, and the overall health of the ecosystem. By continually refining protocols based on these assessments, India's AR initiative remains dynamic and adaptive, ensuring sustained positive outcomes.

Conclusion

The research and initiatives conducted by the ICAR-Central Marine Fisheries Research Institute (CMFRI) in India collectively represent all-inclusive and interdisciplinary approach to address the critical challenges in marine ecology, climate change resilience, and sustainable aquaculture practices. The extensive studies on phenology, distribution, and trophodynamics of climate-vulnerable fishes showcase a comprehensive exploration of marine ecosystems, employing advanced molecular techniques and geospatial analysis for understanding of genetic variations, spatial distribution, and ecological interactions. In assessment of the carbon footprint in Indian marine fisheries demonstrates a commitment to sustainable practices, utilizing standardized methodologies to analyse greenhouse gas emissions throughout the fishing process. The significantly lower emissions in India's mechanized fishery sector, compared to the global average, underscore the success of the implemented methodologies and the potential for global adaptation.The pivotal role of mangroves in climate change mitigation is highlighted through the assessment of carbon sequestration efforts, revealing substantial ecological and economic significance. Geospatial assessments of coastal ecosystems offer valuable insights into the dynamic characteristics of wetlands and seagrass coverage. These assessments, coupled with studies on climate change vulnerability, play a pivotal role in informed decision-making and the development of adaptive strategies at regional levels, thereby enhancing resilience to climate change-induced hazards. These comprehensive tools not only facilitate the mapping of habitats but also enable the georeferencing of fish catch and fishing effort data, while simultaneously linking catch information with oceanographic and biochemical parameters. The integration of these approaches contributes to a holistic understanding of coastal environments and supports proactive measures for sustainable resource management. The innovative climate-resilient farming practices introduced by CMFRI, including cage farming, integrated multi-trophic aquaculture, and seaweed culture, demonstrate a commitment to enhancing productivity while promoting sustainability. These practices significantly reduce the carbon footprint, improve water quality, and offer additional income opportunities for farmers. The successful implementation of Artificial Reefs (AR) stands as a testament to sustained efforts in marine conservation. The evolution of AR protocols, from site selection to impact assessment, ensures that these artificial habitats contribute positively to marine biodiversity and sustainable fisheries. The endeavors of ICAR-CMFRI represent a dynamic and adaptable approach to addressing the complexities of marine ecosystems, climate change impacts, and sustainable aquaculture. By integrating cutting-edge technologies, comprehensive methodologies, and community engagement, these research initiatives contribute significantly to global knowledge and best practices in marine science and conservation.

Further Reading

Akhiljith, P.J., Liya, V.B., Rojith, G., Zacharia, P.U., Grinson, G., Ajith, S., Lakshmi, P.M., Sajna, V.H. and Sathianandan, T.V., 2019. Climatic projections of Indian Ocean during 2030, 2050, 2080 with implications on fisheries sector. Journal of Coastal Research, 86(SI), pp.198-208

Bharti, V., Jayasankar, J., George, G., Ambrose, T.V., Augustine, S.K., Sathianandan, T.V. and Shafeeque, M., 2020. Study on Sea Surface Temperature and Chlorophyll-a concentration along the south-west coast of India. IJMS 49(1) : 51-56

George, G., Menon, N.N., Abdulaziz, A., Brewin, R.J., Pranav, P., Gopalakrishnan, A., Mini, K.G., Kuriakose, S., Sathyendranath, S. and Platt, T., 2021. Citizen scientists contribute to real-time monitoring of lake water quality using 3D printed mini Secchi disks. Frontiers in Water, 3, p.662142

Gopalakrishnan, A and Zacharia, P U and George, Grinson (2020) Impact, vulnerability and adaptation strategies for marine fisheries of India. CMFRI Special Publication (135). ICAR - Central Marine Fisheries Research Institute, Kochi

Hamza, F., Valsala, V., Mallissery, A. and George, G., 2021. Climate impacts on the landings of Indian oil sardine over the south-eastern Arabian Sea. Fish and Fisheries, 22(1), pp.175-193

https://www.cmfri.org.in/nicra/

Joseph, D., Liya, V.B., Rojith, G., Zacharia, P.U. and Grinson, G., 2019. Time series analysis of CMIP5 Model and observed sea surface temperature anomaly along Indian Coastal Zones. Journal of Coastal Research, 86(SI), pp.239-247

Joseph, D., Rojith, G., Zacharia, P.U., Sajna, V.H., Akash, S. and George, G., 2021. Spatio-temporal variations of chlorophyll from satellite derived data and CMIP5 models along Indian coastal regions. Journal of Earth System Science, 130, pp.1-12

Pranav, P., Roy, R., Jayaram, C., D'Costa, P.M., Choudhury, S.B., Menon, N.N., Nagamani, P.V., Sathyendranath, S., Abdulaziz, A., Sai, M.S. Sajhunneesa, T., George G. 2021. Seasonality in carbon chemistry of Cochin backwaters. Regional Studies in Marine Science, 46, p.101893

Sajna, V.H., Zacharia, P.U., Liya, V.B., Rojith, G., Somy, K., Joseph, D. and Grinson, G., 2019. Effect of climatic variability on the fishery of Indian oil sardine along Kerala coast. Journal of Coastal Research, 86(SI), pp.184-192

Sathyendranath, S., Abdulaziz, A., Menon, N., George, G., Evers-King, H., Kulk, G., Colwell, R., Jutla, A. and Platt, T., 2020. Building capacity and resilience against diseases transmitted via water under climate perturbations and extreme weather stress. Space Capacity Building in the XXI Century, pp.281-298

Shafeeque, M., Balchand, A.N., Shah, P., George, G., BR, S., Varghese, E., Joseph, A.K., Sathyendranath, S. and Platt, T., 2021. Spatio-temporal variability of chlorophyll-a in response to coastal upwelling and mesoscale eddies in the South Eastern Arabian Sea. International Journal of Remote Sensing, 42(13), pp.4836-4863

Zacharia, P.U., Rojith, G. and Najmudeen, T.M., 2019. ClimFish NICRA Newsletter Vol. 3

Zacharia, P.U., Sajna, V.H., Rojith, G., Roshen, G.N., Joseph, D., Kuriakose, S. and George, G., 2020. Climate change drivers influencing Indian mackerel fishery in south-eastern Arabian Sea off Kerala, India. Indian Journal of Fisheries, 67(3), pp.1-9

4

Fisheries Education for a Sustainable Future

Ravishankar C. N.

ICAR-Central Institute of Fisheries Education, Mumbai – 400 061, Maharashtra

Email: cnrs2000@gmail.com

'Educate to Escape Extinction' could well be the catchphrase for present times! It is no wonder that all United Nations Member States adopted the 2030 Agenda for Sustainable Development in 2015.Within the global partnership, India too is committed to working towards the 17 sustainable development goals (SDGs) to ensure 'peace and prosperity for people and the planet, now and into the future'. It is agreed that poverty and other deprivations must end, inequality be reduced, health and education systems improved, and economic growth spurred even as we control climate change and preserve natural resources. Activities of all the food production systems, including the fisheries sector, are in one way or the other related to all the 17 SDGs, but the fourgoals majorly related to fisheries research and education institutions are 'zero hunger' (SDG 2), 'quality education' (SDG 4), 'responsible consumption and production' (SDG 12), and 'life below water' (SDG 14).

Being the only national Deemed University dedicated to fisheries education, ICAR- Central Institute of Fisheries Education, Mumbai took the lead in this direction by selecting the theme 'Fisheries Education for Sustainable Blue Economy' for the 3rd International Symposium on Aquaculture and Fisheries Education (ISAFE3), a triennial event of the Asian Fisheries Society (AFS), held at ICAR-CIFE, Mumbai, in May 2018. The delegates deliberated on pertinent topics like 'Education for sustainable fisheries and aquaculture', 'Attracting and retaining youth in the fisheries sector', 'Climate change and environmental concerns through fisheries education', Skill development through training and informal education', 'Entrepreneurship development, and 'Course curriculain Asia-Pacific region for better job opportunities', all of which focused on directing action towards achieving the SDGs. Under a special students' session organised during the event, young minds discussed possibilities of valuation of ICAR-CIFE's human capital, importance of gender studies and social science in fisheries education, extent of entrepreneurship among students, aspirations of youth in fishing communities, potential of online courses and to the extent of social media usage.

This event prepared the ground for the syllabus revision by Broad Subject Matter Area (BSMA) Committee under ICAR-Education Division in 2021, and also generated recommendations that were in tune with India's New Education Policy declared later in 2020. Some of the pertinent recommendations that resonate with SDG of 'quality education'and NEP 2020, and are being implemented by ICAR-CIFE are mentioned below.

- Treat education as the fourth pillar of sustainability and find innovative ways *to bridge the urban-rural knowledge & skill divide* through inspired basic and problem oriented applied research, and better connect between R&D and the rural livelihoods.
- Ensure uniformity with enough flexibility of course curricula and minimum standards in fisheries higher education across India via regulatory bodies (NCHAE/FCI) and by *strengthening governance and leadership in AUs* reforms.
- Foster and strengthen collaboration among countries in SAARC and Asia-Pacific region for joint academic programs, greater faculty-student exchanges, collaborative research programs to constantly enhance the quality of education.
- Improve basic infrastructure in Fisheries Colleges and reorient the education from content-centred to learning-centred approach for preparing next generation of SMART fisheries professionals.
- Bridge the employable skill gaps (40-50%) among fisheries professionals by reorienting course curriculum that is still tilted against skill enhancement (both technical and soft skills) producing mainly job seekers (only 3% have become entrepreneurs) with greater involvement of industry and potential employers.
- Technology enhanced learning offer new opportunities to reach the unreached and strengthen the conventional learning systems by harnessing the use of ICT tools for ODL through MOOCs, webinars, video conferencing, mobile Apps, social media, smart classrooms, IoT, AI, etc.
- Climate change is one of the most pressing global concerns and has already begun to impact fisheries and aquaculture. Create awareness as well as introduce courses on climate change and related aspects

in curriculum to address the lack of conceptual clarity on the subject among fisheries academia, take up multi-disciplinary research program and regularly involve all stakeholders.

- Train the millions of school drop-outs through vocational education following ASCI defined QPs and provide gainful employment in fisheries and aquaculture sector. Introduce programs on scuba diving,mangrove based agro-aquaculture, aqua based eco-tourism, disaster management, island fisheries management, coastal zone regulation and management, blue carbon economy, etc.

Quality fisheries education being the major focus of this national deemed university, ICAR-CIFE, once again took the lead by organising a meeting of the Deans of Colleges of Fisheries and Fisheries Universities of the country in April 2023 to discuss implementation of NEP 2020. The Policy, that holds sustainability at its core, envisions a higher education that is multidisciplinary and holistic in order to create thinking, knowledge driven, ethical, responsible, cultured and innovative Indians who can find and create new livelihood opportunities and work towards a sustainable future. Very rightly, it identifies the practices of early stream selection and discipline rigidity in the Indian education system as major bottlenecks that stand in stark contrast with the high flexibility of curricula in the developed countries, a flexibility that has evidently ignited minds and fueled inventions. This flexibility is not just restricted to choices of disciplines and courses, but is extended to who can enroll for the academic programmes and also to the duration of enrollment. It is expected that this flexibility would eventually translate into broader job eligibility criteria. NEP 2020 expects universities to be large spaces with 3000 or more students. There are presently thirty-four universities and colleges offering higher fisheries education in India, with about 5500 students enrolled at any given point of time. ICAR-CIFE, where only Masters and Ph.D. programmes are offered, will need to work out modalities for expansion. Introducing undergraduate education would be most profitable as young minds benefit best in the company of senior students and scholars. In addition, the Policy encourages universities to offer vocational courses as well, because the multi-disciplinary environment of the campus is expected to benefit the students of these courses in a more holistic manner. In addition, rather than just India centric education, the NEP 2020 points towards an international appeal to be built into degree programmes and curricula being offered in India. Introduction of specially designed broad-spectrum disciplines to attract global students would lead to a more vibrant campus and create an innovative environment that would benefit the Indian fisheries sector. National and international linkages of universities are emphasized in the NEP, and student and faculty exchange can be made possible with more globally appealing courses.

At the Deans' Meet held at ICAR-CIFE, Mumbai, the ten major points deliberated upon include (i) gaps between the present system and NEP 2020 guidelines; (ii) duration of B.F.Sc. (3 or 4 years); (iii) possibility of providing multiple-exit option to undergraduate students; (iv) completion of M.F.Sc. course work within two semesters; (v) possibility of making students of four-year undergraduate programmes eligible for PhD; (vi) output based student research (vii) online learning courses (viii) industry partnership for 'studentpreneurship'; (ix) innovative teaching methods; (x) credit transfer, and significant recommendations/decisions were agreed upon to improve the quality of education.

In present times the significance of online learning cannot be ignored. It is the need of the hour that common teaching platforms be developed for all Fisheries Colleges/ Universities. Online learning can be improved by including multimedia elements like podcasts, videos and interactive activities. The different learning needs of students may be taken into account while creating online courses. Students may learn at their own pace and receiving individualised instruction using tools like adaptive learning technologies. Regular assessments on online learning platforms would be beneficial for students to monitor their development and identify areas that require improvement. There should be opportunities for students to interact with each other and the teacher because online learning can occasionally feel lonely. Discussion panels, live video chats and other interactive exercises can be used for this. Providing prompt and meaningful feedback to students can help them stay on track and improve their understanding of the course material. Automatic feedback software or customized feedbacks by the teacher could be integrated to improve the quality of education.

Agricultural education system is based on the belief that research is an integral part of higher education. In line with SDGs and NEP 2020, most fisheries research is directed towards local, national, or global applications, and it is prioritized based on sustainable development goals, relevance to fish farmers, communities, and industry, but it is imperative that the objectives be clearly defined and designed to maximize hands-on experience, possibility of publications and technology generation. However, to improve the quality of research, it is recommended that on completion of the academic year there should be an evaluation to assess the overall student research outputs and to generate ideas for the next year. Rewarding students and faculty members with best outputs provides encouragement and motivation and this is being done in most institutions.

Fisheries education encompasses other SDGs by promoting use of renewable energy, reducing energy and water requirement, hygiene and biosafety, reducing drudgery in aquaculture and fisheries operations, and waste recycling. Much innovation in directed towards solar powered devices for the fisheries sector, minimising greenhouse gas emissions in aquaculture operations, single cell biomass production for carbon sequestration, and products from fish waste, etc. On-campus energy and water saving, cleanliness, enabling environment for differently abled, gender equality, tree plantation, promoting and conserving biodiversity are the important criteria for university ranking, which has directed higher attention towards the SDGs. These issues are being routinely highlighted through curricular and co-curricular activities.

As we strive towards achieving sustainable development goals, education is our only hope and the seats of higher learning must be strengthened with qualified and dedicated faculties, appropriate infrastructure and support systems, so that the young minds are ignited with knowledge and sensitized to sustain life on this unique planet.

Further Reading

https://www.cife.edu.in/

NAHEP Annual Report 2022-23. National Agricultural Higher Education Project, ICAR, New Delhi

National EducationPolicy 2020. Ministry of Human Resource Development, Government of India

Rathore N.S., Venkateshwarlu, G., Khurana, K.L., 2017. Fifth Deans' Committee Report. Agricultural Education Division, Indian Council of Agricultural Research, New Delhi

Restructured and Revised Syllabi of Post-graduate Programmes (Fisheries Science), 2021. Agriculture and Allied Sciences Volume-5, Directorate of Knowledge Management in Agriculture, Indian Council of Agricultural Research, New Delhi ISBN: 978-81-7164-239-7

5

Innovative Research & Advances in Sustainable Fish Genetic Resource Management in India

Uttam Kumar Sarkar

ICAR-National Bureau of Fish Genetic Resources, Lucknow -226 002, Uttar Pradesh
Email: uksarkar1@gmail.com

Introduction

Aquatic resources refer to the vast and diverse group of water-dependent habitats like rivers, streams, floodplains, reservoirs, ponds, estuaries, lagoons, wetlands, rice fields, open oceans and seas. Sustainable aquatic resources are a foundation for human survival and economic development and support all human, animal and plant life. Aquatic resources are vital for their role in food security, nutrition, livelihood generation and climate change mitigation. Aquatic resources provide livelihood options for many in rural areas of the developing world. Foraging for aquatic biodiversity resources such as crabs, prawns, snails, insects, aquatic plants, etc. in paddy field and other water bodies is an important livelihood component. Aquaculture also plays an important role in low-income countries, which produce over 75% of world aquaculture production generating valuable foods, income and employment in support of the development of disadvantaged regions.They are known to form the wealthiest habitat of global biodiversity and provide a wide range of benefits to the people. Global fisheries and aquaculture production are at a record high of 214 million tonnes with about 58.5 million people employed in the primary sector (FAO, 2022). Even in India, the fisheries sector has been growing remarkably.

India has a rich and diverse aquatic resource ranging from deep seas to lakes, ponds, and rivers which harbours about 10% of the global biodiversity in terms of fish and shellfish species. India has a vast coastline of 8118 km with 5.3 lakh square km of continental shelf which mainly supports marine fishing and navigation. India is also blessed with a 1.95 lakh km long river and canal system which supports numerous towns and cities and freshwater capture fisheries, 8.12 lakh hectares of floodplain lakes, 21 lakh hectares of ponds and lakes, 31.5 lakh hectares of reservoirs, 12.4 lakh hectares of brackish water bodies and about 12 lakh hectares of saline areas. The total EEZ area is 2.02 million square km with a total territorial water of 159,265 square km (Table 1) (National Fisheries Policy, 2020).

Fish diversity in India

Aquatic Genetic Resources Information System (AqGRISI) of India, a database developed by ICAR-NBFGR, Lucknow, mandates the documentation of fish genetic resources of the country for conservation and sustainable utilization for prosperity. The database has documented 3,193 native fish species belonging to 254 families, 53 orders and 1036 genera (ICAR-NBFGR, 2023). As per IUCN categorization, a total of 21 species are critically endangered, 81 endangered, 116 vulnerable, 1204 least concern and 219 data deficient species were documented. Apart from finfish resources, nearly 2934 species of crustaceans, 5070 species of molluscs and 765 species of echinoderms are also contributing to the aquatic genetic resources of the country.

Table 5.1: Aquatic resources in India and their utilisation modes

S. No.	Resource type	Resource size	Fish production system
1	Continental Shelf area (million km^2)	0.53	Capture fisheries
2	Exclusive Economic Zone (million km^2)	2.02	Capture fisheries
3	Rivers and canals (million km)	0.195	Capture fisheries
4	Ponds and tanks (million Ha)	2.41	Aquaculture
5	Reservoirs (million Ha)	3.15	Culture-based fisheries/Stock enhancement/Cage culture
6	Floodplains and lakes (million Ha)	0.812	Culture-based fisheries and pen culture
7	Brackish water area (million Ha)	1.24	Aquaculture
8	Beels (million Ha)	1.3	Culture-based fisheries/ Aquaculture
9	Wetland (Inland and coastal) (million Ha)	15.26	Capture and aquaculture
10	Mangrove (km^2)	4975	Livelihood enhancement

Source: National Fisheries Policy (2020); ICAR-NBFGR & Alliance of Biodiversity Alliance and CIAT (2020).

Transformative efforts in conservation and sustainable utilization of AGR

Exploration, cataloguing, and documentation of AGR

Systematic exploratory surveys were conducted to inventory the fish species in various river basins, including the Indus, Ganga, Gagghar, Luni, Banas, Gandak, BurhiGandak, Bagmati, Brahmaputra, Mahanadi, Krishna, Godavari, Cauvery, Sharavathi, Valapattanam, Chandragiri, Chaliyar, and their respective tributaries. Further, surveys extended to the Andaman and Lakshadweep Islands, as well as selected Ramsar sites. These surveys have yielded remarkable results, including the discovery of 49 new fish species, 6 new shrimp species, and 2 new fish parasites. In addition, 9 fish species have been accurately re-described, contributing significantly to AGR of the country. Books on checklists of fishes belonging to biodiversity hotspots of the country, Northeast India, and the Western Ghats.

The major significant policy documents developed include,

- Guidelines for Germplasm Exchange: Provide a framework for the controlled exchange of genetic resources for aquatic species, facilitating collaboration and diversity preservation
- Inland Aquatic Resources of India: Serves as a valuable resource for understanding India's diverse inland aquatic resources, their ecosystems, and their management
- Ecosystem Services Management: Ecosystem services management highlights the importance of maintaining aquatic ecosystems for the benefit of human well-being, including services like water purification, flood regulation, and fisheries

These policy documents play a pivotal role in regulating and managing exotic species introductions, promoting sustainable resource utilization, and safeguarding the ecological balance of India's aquatic environments.

Genomic resources and databases

Genomic resources and databases are powerful tools that contribute to the sustainable management, conservation, and utilization of aquatic genetic resources. These resources enable more precise and effective strategies in aquaculture, fisheries, and conservation efforts, ultimately ensuring the long-term viability and sustainability of fish populations and industries. The establishment of genomic resources and databases, along with the scientific research and data generated, represents a significant leap forward in the sustainable management, conservation, and utilization of Aquatic Genetic Resources. These tools and information are invaluable for researchers, policymakers, and stakeholders involved in fisheries, aquaculture, and aquatic conservation, contributing to the responsible and informed management of India's aquatic genetic wealth.

Aquatic Genetic Resource Information System of India (AGRISI)

The Aquatic Genetic Resource Information System of India (AGRISI) is an information system on aquatic genetic resources of India and has been launched by the bureau. It provides country-specific information on fish genetic resources as required by the Biological Diversity Act, 2002. It also supports FAO's Report on the State of the World's Aquatic Genetic Resources for Food and Agriculture. AGRISI is a unique platform presently covering >3100 native fish species of India. The system provides information on systematics, biology, distribution, nutrition, and other characteristics. Further, it also includes information on museum specimens and accessions from different repositories. These include data on germplasm and cell lines. Further, AGRISI also links to other molecular resources developed under the National Agricultural Bioinformatics Grid including Fish Barcode Information System (FBIS), a database of hypoxia-responsive genes (HRGFish), a chromosome database of fishes and other aquatic organisms (FishKaryome), fish and shellfish microsatellite database (FishMicrosat) and fish mitogenome resources (FMiR).

The Fish Barcode Information System (FBIS) for fishes of the Indian subcontinent has been conceived and developed by the bureau. It is a platform to assist and manage the acquisition, storage, analysis and exploration of DNA barcode records of the fishes and other aquatic organisms, for species identification, resolving taxonomic ambiguities, and molecular genetic analysis. Species-specific DNA barcodes were developed for around 600 Indian fish species for accurate identification and fisheries resource management purposes. Various legal disputes were resolved based on the DNA barcoding, viz., forensic identification of magur (*Clarias magur*), pomfret (*Pampus chinensis*), the endangered and wildlife-protected whale shark (*Rhyncodon typus*), and sea cow (*Dugong dugon*) using DNA barcoding approach.

Comprehensive Genomics databases, namely FishMicrosat, FMiR, Fish Karyome, and HRGFish with integrated tools for taxonomy, molecular phylogeny analysis, and primer designing for SSR markers were developed and put in the public domain for the utility of various stakeholders. Cytogenetic profiling of 78 endangered and endemic freshwater fish species was generated. The genome size of around 50 fish species was determined by using flow cytometry. Population genetic structure for 30 finfish and shellfish species was ascertained using molecular markers. Microsatellite markers were developed for over 50 important Indian fish species for genetic variation and differentiation studies. Deciphered complete mitochondrial genomes of 28 fish species. Decoded the genomes of various fish species like *Clarias magur, Labeo rohita, Tenualosa ilisha.*

In-situ conservation

The in-situ conservation programmes cover protected areas and fish sanctuaries, the traits of prioritized species, the evaluation of the impact of exotic fishes on natural populations, the development of assays for genotoxicity, etc. Such conservation efforts cannot be meaningful without people's participation through mass awareness programmes. Local bodies and fishing communities must be sensitized about conservation programmes.

Protected Areas (PAs) and Fish sanctuaries

Protected areas (PAs) like national parks, marine parks, biodiversity hotspots, rich biodiverse areas, Ramsar sites, and closed areas play an important role in conserving genetic, species and ecosystem diversity, and in ensuring the delivery of ecosystem services from natural habitats.

Freshwater Aquatic Sanctuary: The establishment of PAs play a key role in the freshwater fish diversity conservation in India (Sarkar *et al.*, 2008). The freshwater habitats located within the terrestrial/wildlife PAs hold rich fish diversity with increased abundance. These areas shall be used/designated as a freshwater aquatic sanctuary (FAS) to conserve the threatened fish species from various threat factors (Sarkar *et al.*, 2013). ICAR-NBFGR has also evaluated the role of terrestrial PAs on fish diversity and community structure in the Gerua River in the Ganga Basin, Pranhita river in the Godavari Basin.

Novel "State fish" concept

The Institute has introduced the concept of 'State Fish of India' to prioritize the species for conservation research and propagation through stakeholder involvement in each state in 2006. This concept serves several essential purposes in the context of aquatic biodiversity conservation and till now 22 states have adopted the concept of state fish in India (Fig. 1). The conceptualization of the 'State Fish of India' is a significant step in the direction of promoting aquatic biodiversity conservation at the state level. It emphasizes the importance of recognizing and protecting region-specific fish species and encourages a collective approach to safeguarding India's aquatic ecosystems and their invaluable genetic resources.

Captive propagation of threatened fishes

Captive breeding programmes are a principal tool to compensate for declining fish populations and supplement and enhance the yields of wild fisheries. Culture, breeding and larval rearing technologies have been developed for various food fishes, ornamental fishes and shrimps for species conservation and sustainable utilization. This also includes the commercially important threatened fishes viz., *Chitala chitala, Ompok bimaculatus* and *Horabagrus brachysoma.*

Linking aquatic germplasm conservation and livelihood: Community Aquaculture

The inventory and cataloguing of aquatic biota are a must for diversifying the germplasm resources of the country. The development of breeding and rearing protocols for such newly discovered species not only contributes to their conservation but also opens up the avenue for their sustainable utilisation. Further, the aquaculture of such species will alleviate the poverty in the local community by uplifting their socio-economic conditions. The Institute has taken initiatives to catalogue and design concepts to validate a replicable working model for harmonising biodiversity conservation and promotion of livelihoods through Community aquaculture in coastal Maharashtra, Pichavaram mangrove region, Tamil Nadu and Lakshadweep islands. Through this program periodical biodiversity surveys have been undertaken to reveal the hidden diversity and captive propagation of such species mitigate the destruction of marine bio-resources and maintain the ecological balance. Furthermore, hatchery production, adaptation, and supply of marine ornamentals by coastal and island communities will create more employment opportunities in this region and raise the hope of the people and their living standard (Kumar *et al.*, 2020).

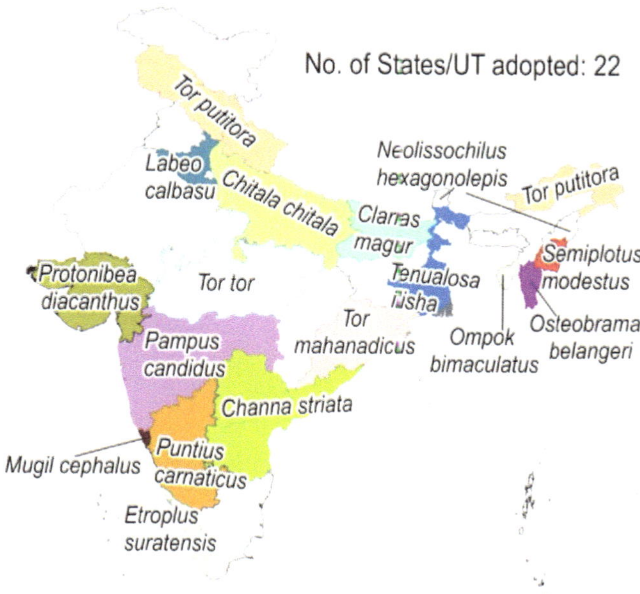

Fig. 5.1: Status of 'State Fish' concept adoption in India

Ex-situ conservation

Fish sperm cryopreservation

Cryopreservation of fish sperm is a promising and very useful technique to facilitate artificial breeding for quality seed production in hatcheries and the development of desirable gene pools. Species-specific sperm cryopreservation protocols have been developed for a diverse range of 39 species. The successful production of viable hatchlings using this cryopreserved sperm has been demonstrated for 24 of these species. The Institute has demonstrated seed production of Indian Major Carps (IMC) using cryopreserved milt in 38 hatcheries in 11 States in the country.

Tissue Voucher Repository

DNA and tissue Banks serve as an efficient mode of storing genetic material for long-term conservation and preservation. Tissue repository accessions are being made with an emphasis on the endemic fish resources of hot spot areas such as the Western Ghats and the northeastern states. DNA and tissue repository at the National Fish Museum building houses over 19000 tissue accessions belonging to commercially/prioritized fish species. The preserved samples include muscles, blood, fin, and gills which are stored in 95% ethanol at 4° Celsius and can be a source of total cellular DNA.

Aquatic Live Fish Germplasm Resource Centre / Live fish gene banks

The Bureau has established Live Fish Germplasm Resource Centre in various parts of the country covering the diverse eco-regions viz., Lucknow, Nagarjuna Sagar in Telangana, Kochi in Kerala, Airoli in Maharashtra, Pichavaram in Tamil Nadu, Guwahati in Assam and Lakshadweep Islands. These centres house the endangered and endemic fish resources of the particular geographical region and research is in progress to develop the Captive propagation protocols for stock replenishment and livelihood development. Simultaneously, they serve as hubs for research, education, and economic development, unleashing their potential to wield a substantial influence on the Indian fisheries and aquaculture industry. This initiative stands poised to make significant contributions to the realms of environmental stewardship and food security.

Quality Fish Seed Ranching

The Institute undertook several captive propagations assisted ranching of commercially important and threatened fish species (IMCs, *Horabagrus nigricollaris* and *H. brachysoma*) in Ganga and Chalakudy Rivers, respectively. Adhering to the conservation of natural genetic diversity, the wild-type broodstock collected from the native environment was then raised and maintained in the Live Fish Germplasm Resource Centres. These brood stocks were then used to produce good-quality fish seeds. To promulgate the conservation in place, the Institute supplies advanced fingerlings to state fisheries departments for ranching in the natural water bodies.

National Repository of Fish Cell Lines (NRFC)

The National Repository of Fish Cell Lines (NRFC), was established in 2010, to maintain, develop and distribute authenticated fish cell lines. Today, NRFC has the world's largest collection of fish cell lines with 81cell line accessions representing 39 fish species. The Institute has developed and deposited 44 of these celllines with 19 being contributed by various organizations. The repository has processed and distributed more than130 mycoplasma-free well-characterized fish cell lines to researchers all over India for R&D works.

National Fish Museum and Repository

ICAR-NBFGR, Lucknow is recognized and designated by the National Biodiversity Authority (NBA), Government of India, as the nodal repository agency for the transfer of fish resources under section 39 of the Biological Diversity Act, 2002 of India. The institute is strengthening this program by establishing the National Fish Museum and Repository. The full-fledged museum facility was dedicated to the nation on the 14thof April, 2023 by the Hon'ble Secretary, DARE, and Director General, ICAR, Dr. Himanshu

Pathak which displays finfish and shellfish voucher specimens of freshwater, marine, and brackish water environments for research and education of students, teachers, scientists, and the general public. The museum also holds a radiographic facility to comprehensively understand fish vertebrae counts, fin rays, and other osteological features (Fig. 2).

Exotics and Quarantine

This institute undertakes periodical surveys in the natural aquatic ecosystems of the country to assess the spread and abundance of various exotic fishes. Documented exotic fish species such as tilapia, common carp, African catfish, red piranha, pacu, and sucker catfish in various aquatic ecosystems in the country. The major significant policy documents developed by the bureau include National Strategic Plan on Aquatic Exotics and Quarantine: Outlines the national strategy for the management and quarantine of exotic aquatic species, ensuring their safe introduction and containment and National Exotics and Quarantine Guidelines: Offer comprehensive protocols for the introduction, management, and quarantine of exotic species in aquatic ecosystems.

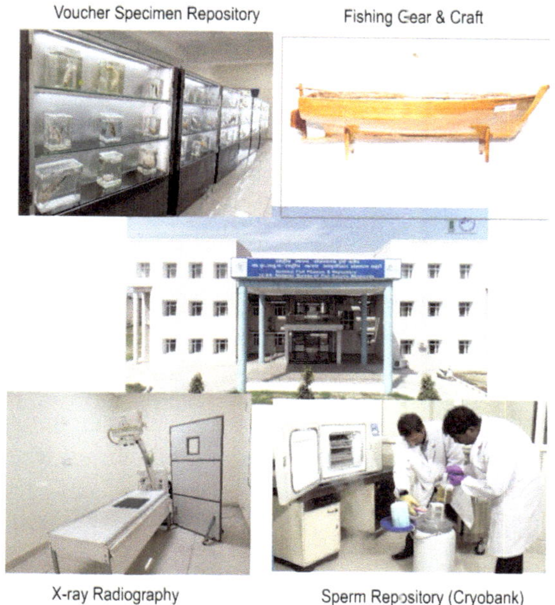

Fig. 5.2: Components of National Fish Museum and Repository of ICAR-NBFGR, Lucknow

Aquatic animal health management

The loss of fish species in different biodiversity-rich natural waterbodies is potentially inked with various natural or anthropogenic impacts or many a time a multiple combination of both. Among these disease outbreaks is one of the main reasons for the decline in the stock of particular fish or alteration of population structure and diversity of different fish species in the wild especially those that are threatened and endemic. Some of the sought-after groups of pathogens responsible for this are viruses, oomycetes and parasites especially of transboundary and emerging in nature. These pathogens can not only be held responsible for biodiversity loss in the wild but also result in cross-species transmission and thus increased fitness in newer environments and different geographical locations. The Institute is undertaking a mega pan India passive disease surveillance program covering 19 states with the help of 31 collaborating centres including line departments of various states. The program reported nine fish pathogens from the country for the first time. The initiative also strengthened the network of diagnostic laboratories across the country and has brought global recognition in the area of fish disease reporting. "ReportFishDisease(RFD)", an online app for a farmer-based disease reporting system for improving the reporting of aquatic animal diseases in the country has been developed by the Institute. The app has been developed National Surveillance Programme for Aquatic Animal Diseases (Phase- II), funded under Pradhan Mantri Matsya Sampada Yojana (PMMSSY) by the Department of Fisheries, Ministry of Fisheries, Animal Husbandry and Dairying, Government of India. Using the app, the farmers can report the incidence of disease in finfish, shrimps and molluscs on their farms with the field level officers and fish disease experts as and when required and get scientific advice for quickly addressing the disease problem in their farms. With the help of this Farmer-First & Citizen-Science Initiative, the economic loss due to disease outbreaks will be reduced to a greater extent, thereby increasing fish farmers' income.

Way forward

Modern advanced technologies would be of immense use for AqGR conservation and management. AI (Artificial Intelligence) in fisheries resource management could be utilized in the area of data collection and analysis. Unlike traditional data collection and analysis, AI-powered tools- can automate the process and provide real-time data on fish populations, their distribution, and their behavior for effective fisheries management plans. Further, various AI algorithms could able to analyze large volumes of data and identify patterns and trends, for example, they can detect changes in fish migration patterns or identify areas where illegal fishing activities are taking place. In addition, AI tools enable fisheries managers to respond quickly and effectively to emerging threats and take appropriate action. AI can also play a crucial role in improving the efficiency of fisheries operations using drones. Similarly, the eDNA-based survey will aid in cataloguing the AqGR in difficult areas and exotic fish species management (Fig. 3).

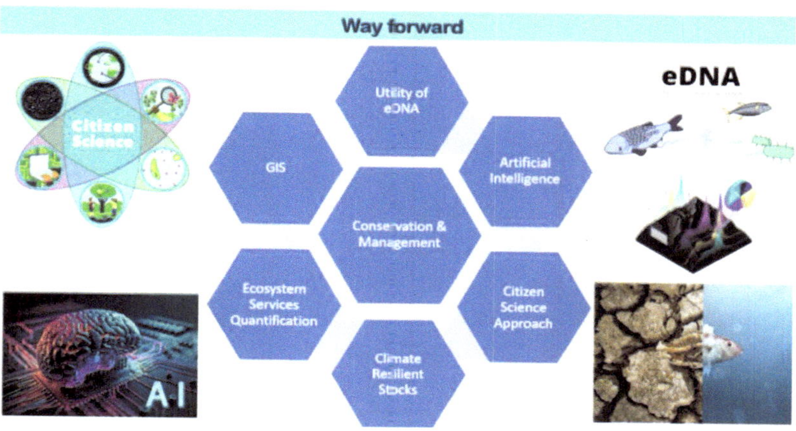

Fig. 5.3: Innovative approaches for conservation and management AqGR

Data collection for devising management of AqGR is a cost, time and manpower oriented approach. One easier avenue for acquiring such data involves harnessing the growing momentum of citizen science, which is defined as a form of open collaboration where individuals willingly participate in the scientific process. If properly planned and implemented, citizen science has the potential to serve as a vital collaborative approach to address critical data gaps in fisheries science and management. This entails complementing existing data collection programs with new projects collaboratively devised and executed by both scientists and volunteers, including fishermen. This collaborative effort holds the promise of filling knowledge gaps with valuable information that could significantly contribute to informed fisheries management. By involving multiple stakeholders in the scientific aspects influencing management decisions, this approach could enhance the decision-making process.

Furthermore, it has the potential to foster trust and transparency among fishers, scientists, and managers. This collaborative model ensures that a broader range of perspectives and expertise is brought to the table, creating a more robust foundation for effective fisheries management. Climate change impacts the AqGR and studies are warranted for identifying climate resilient stocks for a sustainable future for fisheries and associated ecosystems.

Conclusion

Aquatic resources are vital for human survival, and economic development, and to support all human, animal and plant life. Increased anthropogenic activities coupled with climate change have been degrading aquatic resources. Sustainable development and use of aquatic resources are the need of the hour considering their role in food security, nutrition, ecosystem services and livelihood generation. India is blessed with vast aquatic resources that need to be harnessed sustainably and efficiently to meet the nutritional security and needs of our growing population. The insights into institutional transformative efforts for aquatic genetic resources illuminate the substantial progress achieved in the domains of fisheries and aquaculture. These initiatives have played a pivotal role in addressing the existing status, the multifaceted challenges associated with the management, and conservation of aquatic genetic resources. The trajectory of the bureau underscores its unwavering commitment to fostering innovation, conducting cutting-edge research, and preserving the genetic diversity of aquatic species. More than a pursuit of scientific excellence, these endeavours hold the promise of forging a future that is sustainable, food-secure, and ecologically responsible. As this journey continues, the valuable lessons derived from experiences offer guiding principles for institutions across the globe, all striving to ensure a sustainable and prosperous future for aquatic genetic resources. The emerging areas such as citizen science, AI and machine learning, eDNA and advanced GIS tools would be aiding in the AqGR conservation and management of the country.

References

Draft National Fisheries Policy, 2020. National Fisheries Development Board. Ministry of Fisheries, Animal Husbandry and Dairying, Govt. of India, New Delhi.

FAO. 2022. The State of World Fisheries and Aquaculture 2022. Towards Blue Transformation. Rome, FAO. https://doi.org/10.4060/cc0461en

Glaser, M., Glaeser, B. (2012). The social dimension of social-ecological management in Treatise on Estuarine and Coastal Science, eds E. Wolanski and D. McLusky (Waltham, MA: Academic Press), 5-30.

GoI (2010). Draft Guidelines for Integrated Water Resources Development and Management. Central Water Commission, Ministry of Water Resources, RD&GR, Government of India (GoI), New Delhi.

ICAR-NBFGR & Alliance of Biodiversity Alliance and CIAT 2020. Inland Aquatic Resources of India and its Impact on the Ecosystem Services.

Kumar, T.T.A., Charan, R., Jayakumar, T., Tyagi, L.K., Saravanane, N., Mohindra, V., Hisham, J., Lal, K.K. (2020). Framework for participatory linkage of marine ornamentals germplasm conservation to livelihoods: Is community aquaculture an inclusive option? Aquaculture Asia, 24(4), 3-9.

Sarkar, U.K., Pathak, A.K., Lakra, W.S. (2008). Conservation of freshwater fish resources of India: new approaches, assessment and challenges. Biodiversity and Conservation, 17(10), 2495-2511.

Sarkar, U.K., Pathak, A.K., Tyagi, L.K., Srivastava, S.M., Singh, S.P., Dubey, V. K. (2013). Biodiversity of freshwater fish of a protected river in India: comparison with unprotected habitat. Revista de biologia tropical, 61(1), 161-172.

6

Sustainable Shrimp Aquaculture through Diversification and Genetic Improvement of Native Species of Shrimp

Akshaya Panigrahi* and Kuldeep K. Lal

ICAR-Central Institute of Brackishwater Aquaculture, 75, Santhome Highroad, MRC Nagar Chennai, India – 600 028

**Email: apanigrahi2k@gmail.com/akshaya.panigrahi@icar.gov.in*

Introduction

With an annual production of 6.5 million tons worldwide, penaeid shrimp remains one of the most traded seafood commodities across the world (FAO, 2020). India, one of the top producers of shrimp contributed to an export market of 0.84 million tons, or USD $5.8 billion and the market is projected to expand at a CAGR of 9.60% from 2023 to 2028, and it is anticipated to reach a volume of around 1.47 million tonnes. Native penaeid shrimp like Indian white shrimp (*Penaeus indicus*) and black tiger shrimp (*Penaeus monodon*) initially dominated Indian shrimp aquaculture. However, the growth of native species was hindered due to viral diseases such as white spot syndrome virus (WSSV) which led to the introduction of genetically enhanced exotic specific pathogen free (SPF) shrimps, *Penaeus vannamei*. Due to the reliance of shrimp farming on a single exotic species, the industry is facing difficulties.

Due to inbreeding, poor seed quality, and emerging diseases, it is highly risky to depend on one species for the production of 10 lakh tons with huge investments on farming infrastructure and the livelihoods of two lakh farm families directly and several lakhs' families indirectly associated in the ancillary sectors. Indian white shrimp *P. indicus* is an important candidate species and can be deliberated as an alternative species to both the *P. vannamei* and *P. monodon*. It is having good potential for higher growth through genetic selection and adaptability for high-density culture in coastal waters. At the moment, *Penaeus vannamei*, a non-native species, is the sole species used in Indian shrimp farming. It accounts for over 90% of shrimp produced in India and over 76% of shrimp produced worldwide (FAO, 2019). This was made feasible by the introduction of genetically modified animals that had better features for aquaculture and biological benefits, like column feeding patterns, ease of adaptation to high density culture systems, and relative ease of captive breeding. In addition, the exotic *vannamei* has grown to be the largest threat to the Indian shrimp culture sector due to a number of issues with larviculture (Zoea 2 syndrome), hatcheries (deterioration of male reproductive quality), and the production system (various current and emerging diseases such as hepatopancreatic microsporidiosis, growth retardation, poor survival, and running mortality syndrome). The over-reliance on a single species combined with the theory of an inbreeding depression is shown to be detrimental to the prawn industry's ability to expand. It is determined that the growth of native shrimps could be a feasible alternative for the industry's long-term survival in this setting. None of the candidate species of Indian penaeids has a domestication program in India. The creation of genetically improved broodstock through selective breeding is a pressing necessity, though laborious and time-consuming process. *Penaeus indicus*, the Indian white shrimp, seems to be the best substitute species in this regard (Vijayan *et al.*, 2019; Panigrahi *et al.*, 2020a).

ICAR-CIBA initiatives

Shrimp farmers were obliged to explore for alternative tactics after experiencing repeated crop losses and a constant decline in yield, despite the early success of *P. vannamei* culture. At this stage, ICAR-CIBA held stakeholder meetings and presented the idea of implementing better breeding and culture techniques for the native *P. indicus* species. After multiple workshops and awareness campaigns, and after seeing the potential of this species, a number of farmers stepped forward to support it. Several maritime states (Odisha, West Bengal, Andhra Pradesh, Tamil Nadu, Kerala, and Gujarat) have successfully experimented with desi prawn farming, thanks to internal financial support from the National Fisheries Development Board (NFDB). In keeping with the goals of the "Make in India" project, these front-line demonstrations through partnership farming were conducted with the intention of spreading improved breeding and culture techniques and promoting this native species as a supplementary species.Farmers were given disease-free quality seeds to stock in their own ponds. Subsequently, regular technical suggestions were offered throughout the culture practice. The results were excellent at the majority of the demonstration locations, with attainable production levels of 3-8 tonnes/ ha and survival rates as high as 98%. When compared, Indian white shrimp performed similarly to Pacific white shrimp in terms of development rate, final Average Body Weight (ABW), adaptability to diverse farming systems, crowding density, productivity, and profitability.

Fig. 6.1: Indian white shrimp (*P. indicus*) and its propagation in different states

Furthermore, the institute has carried out research projects to delineate genetic stock, as well as whole genome sequencing, which is further refining the aquaculture of *P. indicus* by giving scientific data that is crucial for genetic development.

Key Results/Insights

Multi-location *P. indicus* culture demonstrations conducted nationwide could generate real-world data on production performance, productivity, and disease prevalence across various farming systems. All demonstration trials employed post larvae produced by WSSV-free broodstock (Anand *et al.*, 2019). Suitable *P. indicus* culture technology for extensive and intensive farming is created with SOP and significant BMPs to achieve maximum production (Panigrahi *et al.*, 2020a; Lalramchhani *et al.*, 2019; Antony *et al.*, 2019) and minimize hazards. High density culture standardisation and demonstration employing innovative culture procedures (Panigrahi *et al.*, 2020b) are done with the construction of an Indian white shrimp database.

On this multi-location culture trial with low and high-density farming models (15 to 45 nos./m^2), carried out for 80 to 125 days of culture, an ABW of up to 24-25 g, survival up to 98%, and a production range of 1.5 to 7.0 t/ha were effectively achieved with good economic gain. For example, in Odisha, average production of 1142 and 4418 kg/ha with ABW of 28-30 g and 17-20 g were produced at lower (10 PL/m^2) and higher stocking densities (35 PL/m^2), respectively. The shrimps were fed CIBA-formulated feed indicus^{plus}. The shrimp were sold at the rate of Rs 330 to 420/kg against a production cost of Rs 230/kg. Similarly, as part of the research, almost 250 lakh shrimp seeds were dispersed to 25 farm locations across India under various salinity regimes and farming systems, using modified location-specific culture methods.

Average daily growth (ADG) was found to be 0.195±0.032 in the salinity range of 15 to 35 ppt, which was significantly greater than the low ADG of 0.125 ±0.021 in the very low and very high salinity ranges (<7 ppt & >40 ppt). Over the course of 80–120 days, two distinct stocking densities (low and high) and salinity regimes (3–60 ppt) were used to evaluate the growth performance and health status. A 20–30% increase in stocking density resulted in a significant ($P<0.05$) increase in total productivity (kg/ha), which was similar across salinity ranges.We decoded the entire mitochondrial DNA genome of *P. indicus* for the first time, filling gaps in existing genetic data on cultivated shrimp species. The genome sequence, together with annotations, has been submitted in GenBank under accession number KX462904. Furthermore, evaluating diverse stocks throughout the Indian coast and conducting divergence research will aid in developing a base population for the genetic improvement program of this native species.

Important characteristics like "growth" and other criteria will be included in the selection program going forward in order to achieve the appropriate heritability and genetic gain. After substantial stock improvement in the G6/ G7 generations, when they can be hired in BMC for additional propagation, the selection process will be upheld. As an early project spinoff, commercial companies may be granted access to a portion of the resulting stock for the manufacturing of SPF PL of *P. indicus*. This is what will drive the development of this significant native shrimp's sustainable aquaculture.

ICAR-CIBA cleaned broodstock bred *P. indicus* shrimp seeds was given to shrimp farmers to assess its production potential at farmers ponds under "Genetic Improvement Programme of *Penaeus indicus*". In this connection an on-farm harvest of Indian white shrimp indicus was held on 10-12-23 under the Matsya Sampada Jagrukta Abhiyan scheme at Utukuru village, Nellore district in Andhra Pradesh. It was recorded that Indian white shrimp was refractory to white fecal disease in the culture ponds which is the devastating problem in shrimp farming as of now. The *P. indicus* shrimp exhibited good growth rate on an average @ 0.3 g/per day and attained 20 g in 90 days of farming with cost effective feed. Mr. Krishna Reddy progressive farmer, who farmed the *P. indicus* shrimp expressed his satisfaction on the performance of *P. indicus* shrimp. The farmer stocked 1.6 lakhs *P. indicus* seeds in 1.7 acre pond and the shrimp could reach a weight of 20 g in 93 days with much lesser production cost.

Step towards genomic selection

ICAR-CIBA have decoded the complete genome of this shrimp species which is found to have the following characteristics

- 1.93 Gb length (78% coverage)
- 44 scaffolds are longer than 5 Mb length (synonymous to pseudo-chromosomes) 49.31 % repetitive elements
- Highest SSRs (31.99 % among sequenced animal genomes)

We now have the knowledge of full set of genes and the pathways governing the species. We have also generated full-length isoform-level transcript sequence information for *P. indicus*. This information would help in deeper functional understanding of target traits and subsequently genomic selection.

Future Roadmap of Indian white shrimp

Pilot-scale shrimp breeding and domestication facility which includes development of infrastructures like primary and secondary quarantine units, nucleus breeding centre and founder population stock of *P. indicus* are being established to initiate genetic improvement program in the initial phase. The selection program will continue with important traits like 'growth' and other parameters to obtain the desired heritability and genetic gain. The selection process will be continued till G6/G7 generations with substantial improvement in the stock when they can be recruited in broodstock multiplication centre (BMC) for further propagation. A portion of the resulting stock can be shared to private players for production of SPF PL of *P. indicus* as an immediate spinoff of the project. Establishment of BMC for further propagation of the brooders and catering to the industry has to be executed. The broodstock multiplication centers will be established at selected sites and multiple generation lines will be developed. This work will be taken in a consortium mode with CIBA as a nodal agency, along with technical support from consortium partners such as the Ministry of Fisheries, Government of India, MPEDA, NFDB, and private entrepreneurs. In the subsequent phases the selection program will be continued to obtain the desired heritability and genetic gain.

References

Ashok Kumar J, Vinaya Kumar K, Panigrahi A,Gangaraj KP, Suganya N, Raymond Jani Angel J and Shashi Shekhar M (2022). Hepatopancreas Transcriptome and Gut Microbiome Resources for *Penaeus indicus* Juveniles. Frontiers Marine Science.8: 809720.

Panigrahi A, Esakkiraj P, Das RR, Saranya C, Vinay TN, Otta SK, Shekhar MS (2021). Bioaugmentation of biofloc system with enzymatic bacterial strains for high health and production performance of *Penaeus indicus*. Scientific reports. 11(1):1-3.

Panigrahi A, Das RR, Sundaram M, Sivakumar MR, Jannathulla R, Lalramchhani C, Antony J, Shyne Anand PS, Vinay Kumar K, Jayanthi M and Dayal JS, Cellular and molecular immune response and production performance of Indian white shrimp *Penaeus indicus* (H. Milne-Edwards, 1837), reared in a biofloc-based system with different protein levels of feed. Fish & Shellfish Immunology (2021) 119:31-41.

Panigrahi, A., Das, R.R., Sarkar, S. *et al.* Biofloc-based farming of Indian white shrimp, *Penaeus indicus*, in recirculating aquaculture system (RAS) enriched with rotifers as feed supplement. Aquacult Int (2022). https://doi.org/10.1007/s10499-022-01000-8.

Panigrahi, A., L. Christina, Shyne Anand *et al.* 2019. Production Performance of Indian white shrimp, *Penaeus indicus*: Demostrated along the different coastal states of India, ISBN-978-81-945379-1-5

Lalramchhani C, Panigrahi A, Anand PS, Das S, Ghoshal TK, Ambasankar K and Balasubramanian CP (2020) Effect of varying levels of dietary protein on the growth performances of Indian white shrimp *Penaeus indicus* (H. Milne Edwards). Aquaculture 519:734736

Anand PS, Balasubramanian CP, Francis B, Panigrahi A, Aravind R, Das R, Sudheer NS, Rajamanickam S and Vijayan KK (2019). Reproductive Performance of Wild Brooders of Indian White Shrimp, *Penaeus indicus*: Potential and Challenges for Selective Breeding Program. Journal of Coastal Research. 86(sp1):65-72.

Anand PS, Aravind R, Biju IF, Balasubramanian CP, Antony J, Saranya C, Christina L, Rajamanickam S, Panigrahi A, Ambasankar K, Vijayan KK. Nursery rearing of Indian white shrimp, *Penaeus indicus*: Optimization of dietary protein levels and stocking densities under different management regimes. Aquaculture. (2021) 542:736807.

Tomy S, Saikrithi P, James N, Balasubramanian CP, Panigrahi A, Otta SK, Subramoniam T and Ponniah AG. (2016). Serotonin induced changes in the expression of ovarian gene network in the Indian white shrimp, *Penaeus indicus*. Aquaculture. (2016) 452:239-46

Vijayan KK. (2019). Domestication and Genetic Improvement of Indian White Shrimp, *Penaeus indicus*: A Complimentary Native Option to Exotic Pacific White Shrimp, *Penaeus vannamei*. Journal of Coastal Research. 86(sp1):270-6.

7

Innovations in Freshwater Aquaculture in India for Inclusive Growth of the Sector

Pratap Chandra Das* and Himanshu Sekhar Swain

ICAR-Central Institute of Freshwater Aquaculture, Bhubaneswar, India

**Email: pratapcdas@yahoo.com*

Introduction

Recognition of fish as a safe protein in the Indian food basket has given impetus for continued faster growth of the aquaculture sector as compared to other agricultural enterprises. Starting with a home-stead activity in the Eastern Indian States, aquaculture, particularly in the freshwater ponds and tanks, has reached the status of industry in the country. India has been the second largest fish producer from aquaculture in the globe since last several years. Out of the total 16.25 MMT annual fish production (2020-21) of the country with a growth rate of 10.37% in 2021-22 over 2020-21 and 8.61% for the 8-year period from 2014-15 to 2021-22. The inland freshwater sector contributed approximately 74.6% of the total production constituting almost 80% from aquaculture and rest from the capture fisheries.The average fish production from aquaculture ponds have crossed 3.0 t/ha with farmers in many states registering more than 8.0-10.0 t/ha production level. Fish consumption of the fish-eating population has grown significantly to over 13 kg per capita per annum in 2022-23 from 7 kg in 2011-12 (NSS). Being the most populous country in the world and with the growing fish-eating population and, India adds a sizable amount of demand for fish every year and there is a huge domestic market for fish in the country. This provides an enormous scope for increasing the fish production from the inland resources. At present, only 70% of the pond resources in the country are being used for farming while cage and pen culture activities are ongoing in few selected reservoirs only. In that way, availability of huge cultivable aquatic resources in the form of 2.45 million ha ponds and tanks, 1.2 million ha flood plains and 3.15 million ha of reservoirs also further expands the scope of increasing the fish production. India is also biological hub for fish fauna with availability of more than 1000 freshwater fish species, many having culture potential. Last 70 years of research and development activities in the country have also led to the development of standard technologies to utilise almost every type of water bodies for fish production using the different fish species.

Availability of resources, potential fish species, culture technologies, strong extension network and continued fostering of Central and State Governments for the fisheries and aquaculture sector has supported the growth of the aquaculture sector. Despite the fact that there is stagnancy in the marine capture production and brackishwater resources being used mostly for export-oriented shrimp production, the freshwater aquaculture sector has been able till date not only in meeting the ever-increasing fish demand of the country, but also there has been significant increase in the export of the aquatic products. However, with further increase in the fish demand in the future, the sector is expected to face many challenges some of which include water scarcity, climate changes effect on fish breeding and growth, poor mechanization, labour shortage, intra- and inter-sectorial competition for inputs and also the concern for environmental sustainability. Sustainable development also requires aquaculture to make itself more remunerative, produce more fish with environment-friendly protocol, increase species diversification to create options for consumers while preserving the species and ecosystem diversities. The R&D system in the country has been apt in addressing these requirements over the years, several new fish production packages of practices have been developed, adopted practices have been further refined and modified to keep the track of growth upward. Some of the innovations and strategies adopted are described in the following pages.

Innovation in seed production

Successful aquaculture operation largely depends on the large-scale availability of quality seed. Therefore, seed production in controlled conditions is a prime precursor for aquaculture development. Although induced breeding technique with use of the pituitary gland was the most important epoch-making technology of last century for aquaculture, it was often marred with problems of efficiency issues of PG extract, two-time injections and too much handling stress of brooder. However, development of synthetic version of inducing agent has overcome the problems and made the induced breeding easy and widely applicable for different types of fishes. Following success of the Ovaprim, a number of other synthetic inducing agents followed the suit and had revolutionised the carp hatchery operation and large scale seed production. The list includes ovatide (SGnRH-A+pimozide and later SGnRH-

A+domperidone), Wova-FH (SGnRH-A+ domperidone), Gonopro, etc. Higher spawning fecundity, improved fertilisation rate, % hatching and spawn recovery achieved with use of these inducing agents besides increased post-breeding brooder recovery. Development of synthetic inducing agents have also helped in developing induced breeding protocol for many other economically important groups of species such as catfishes, murrels and ornamental fishes. Use of LHRH is also in vogue for the breeding catfishes. Hormonal implant techniques have also been successfully developed for the induced breeding of catfishes and murrel. Today, induced breeding technique has been standardised for more than 40 commercially important freshwater food fishes apart from many ornamental fishes. Multiple spawning, i.e. breeding same fish four times during a season (March, May, July and September), developed by ICAR-CIFA has increased the average spawn productionand demonstrated 2-3 folds higher spawn recovery over conventional single breeding. Similarly, techniques of carp milt cryopreservation reduced requirement of male broodstock population in the hatchery. The technique has also been used for stock upgradation to overcome the inbreeding deficiency in many hatcheries. Off season breeding through environmental manipulation is another tool effectively used for seed production prior to the natural breeding season. Development of the broodstock diet CIFABROOD™ in 2010s, that ensures early broodstock maturation, has further made the early breeding protocol easy and effective.

Parallel to the development of inducing agents, series of hatchery models have been developed over the years. Some important models innovated over time include hapa, earthen pot hatcheries, Tub hatchery, Cemented cistern hatchery, Glass jar hatchery, Galvanised iron jar hatchery, Shirgur's bin hatchery, Circular cistern hatchery, Eco-hatchery and FRP hatchery. Of the above hatchery models, the Eco-hatchery model is continued now days for mass scale carp seed production whereas the jar hatchery system (glass, concrete types) is also in use in some hatchery and operated for low scale of production. The FRP hatchery system developed for carps has also proven to be effective for the breeding of several minor carp species. Further, it has also been suitably modified for breeding of other important species such as magur, singhi, pabda etc.

Innovation in fry and fingerling rearing

Fish seed rearing activity includes two stages rearing from spawn to fry stage in 15-20 days nursery phase for and fry to fingerling stages in 2½ to 3 months rearing phase. Both the stages need specialized protocol for rearing. Conventionally, these seed are reared in prepared earthen pond with rearing density of 5-10 million spawn/ha and 0.2-0.3 million fry/ha in nursery and rearing phase, respectively. However, both fry and fingerling rearing in earthen pond are often marred with problem of poor survival owing to problems of predation (bird, predatory and weed fishes and aquatic insects) with limited control. In recent years, technologies of high density seed rearing in concrete tank system has been developed with reduced feed and water use, but with greater survival and seed production. Additional provision of inputs in the form of aeration and balanced diet further improves seed production in these tank system. Use of concrete tank system is now being promoted to ensure higher seed yield with less use of water. While rearing density of 500 spawn/m^2 is followed in the earthen pond for the Indian major carps with an average survival of 40%, the density could be increased to 2000 spawn/m^3. Water depth in such system could be suitable reduced to 0.75 m while use of supplementary feed reduced by 25% with resultant higher survival upto 60% in nursery and 75% in fingerling raising. Seed rearing in biofloc system has emerged as a potential intervention in recent days. The rearing density of spawn could be increased to 6000/m^3 in biofloc system with survival as high as 85-90% after 20-25 days. In-situ live food in terms of bioflocs in such system not only ensures continuous food availability, but also helps in size uniformity of the seed. High density fingerling rearing of carps (major and minor carps) and catfishes (magur, singhi, stripped catfishes) in all these tank systems (concrete tanks, biofloc system) have also reported to yield higher growth and survival.

Production of stunted seed is another trend observed in recent years as such seed which are usually bigger in size and available round the year help the grow-out farmers to practice continuous culture in perennial ponds, especially those practice multiple stock-multiple harvest cropping pattern. The stunted seed are reared at high density with sub-optimal feeding to restrict their growth. Popularization of the use of stunted seed has created a third tier of seed farmers besides the fry rearer and fingerling rearer. Stunted seed production has high economic viability and has been adopted by some farmers as full-time profession.

Strategies to increase fish production

a) **Effective technology dissemination for scientific fish farming:** Growth of fish is a biological process and it requires sound basic knowledge and continuous attention for management of fish culture system with regard to environmental health, nutrition and disease management. Ponds, tanks and flood plain resources form the major resources for freshwater aquaculture in India. However, these resources are sparsely distributed which has been the major bottleneck for technology percolation to the fields. However, training and skill development in fish farming has been continuously emphasized and implemented through the extension system over the years. The strong extension network of Fisheries Departments in all States, 731 KVKs and NGOs in the country, besides Research Institutes, have been able to improve the awareness on fish farming among farmers in general. In recent years, implementation of several Central and State sponsored schemes for development of aquaculture sector with targeted beneficiaries have helped in effective translation of the scientific fish farming in to the farmers' fields, that have not only had its multiplier effect, but also increased the knowledge level of average Indian fish farmers.

b) **Resource renovation and horizontal expansion:** Composite fish farming of the three Indian major carps; catla, rohu and mrigal alone or along with the three exotic silver carp, grass carp and common carp has been the main practice followed till date in Indian pond farming. In 2021-22, the group carps constituting major carps and exotic carps, contributed 69.93% of the inland production of the country that signifies the commercial importance of the group. Since the surface and bottom area of the carp ponds are constant limiting the density of the surface and bottom feeders, increased productivity of the pond can be expected only through provision of greater depth for the column feeder (rohu). Therefore, 1.5-2.5 depth is considered a suitable range for grow-out farming of carps. This fact is often ignored by the farmers and carps are being cultured irrespective of the water depth often leading to poor production. While wide range of technologies are available for production of seed and grow-out culture of varied species, identification of appropriate technology should largely base on resource characteristic for its effective utilisation. For example, a shallow pond can be more effectively used for seed production compared to economic return from grow-out farming. Further, development of culture technologies for many high valued species such as minor carps, catfishes (magur, singhi, pabda), murrels, scampi and other non-carp species offer opportunity for suitable selection technologies as per resource characteristic. In recent years, renovation of the ponds and tanks has been an integral part in almost all the promotional schemes for implementing the scientific fish farming, that has helped in increasing the yield level from average pond. Further, continuous awareness and promotional schemes in various subsectors have encouraged farmers to adopt resource specific intervention, i.e., hatchery, fry, fingerling and juvenile rearing to get better return.

Of the 2.45 million ha pond resources available in the country, almost 70% is being utilised at present. While efforts have been made over the years through various development scheme to bring these ponds under scientific farming, creation of new water bodies (ponds and tanks) has been another major area of promotion from the government for increasing fish production and improving the water harvesting potential. Short-term leasing policy of community ponds and small reservoirs is always disadvantageous for the resource since the leasee would always try to exploit the maximum benefit out of it without giving a thought for its maintenance and long-term preservation. Such leasing policy have been suitably modified in many states to provide long duration lease holding which has promoted horizontal expansion of culture area under scientific fish farming.

c) **Vertical increase in productivity** An increase in the pond productivity is achieved through effective management of the pond environment, judicious management of inputs and taking care of the nutrition and health aspects of the cultured fish. Over the years, the grow-out fish production technologies have been refined and modified for various production systems and cultured species. Increasing species spectrum, devising suitable species combination, changing the cropping pattern, breed improvement and use of quality fish seed with higher growth and disease resistant traits, etc. are some of the means employed with proven enhanced yield.

i. ***Species diversification:*** Need to produce more fish to cater the increasing demand in the country has not only led to increase the intensity of the farming operation in recent years, but also many new species have been added to the culture system. Species diversification in culture system has shown good dividends of improved yield and farm income, efficient input use and production of varied fish protein, besides diversity conservation through culture propagation. Research efforts have made possible to bring many potential minor carps, high valued catfishes, air-breathers into the loop of culture species spectrum. Though the Indian major carps and the exotic carps are continuing to contribute significantly to the inland production, the production of the later has gone down in recent years. Striped catfish has become the next important cultured species in the culture system, specifically in Andhra Pradesh, Chhattisgarh, Jharkhand, Bihar and Uttar Pradesh. From 2005 onwards, culture area for the striped catfish increased in Andhra Pradesh upto 32,000 ha in 2010 with an average yield of 17-20 t/ha, while the yield level had often reached to 40 t/ha with intensive monoculture set up. But its culture got a setback due to the lower market price. Pacu is yet to be approved as a cultivable species in the country. But its culture area has increased during last 10 years, particularly in Andhra Pradesh with present coverage of more than 2500 ha. There has been a growing interest for culture of improved variety of mono-sex tilapia and red tilapia which have received nod as cultivable species with certain restriction. While pacu is being cultured mostly as a component species in carp polyculture system, tilapia and stinging catfish (*Heteropneustes fossilis*) have been widely adopted as species of choice for the bio-floc system. Monoculture of magur (*Clarias magur*) and butter catfish (*Ompok pabda, O. bimaculatus*) are increasing popularity in the North Eastern states with 1.0-1.5 t/ha production range in both species. Culture of murrels is on rise in West Bengal and south Indian states, but yet to flourish due to non-availability of seed in large scale. Recent success in the mass scale seed production technology of *Channa striata* at ICAR-CIFA has opened scope for its expansion in the culture system. Similarly, intervention in the form of breed improvement programme in freshwater prawn *Macrobrachium rosenbergii* at ICAR-CIFA resulting in 30% higher weight gain after 12 generation has started reviving the freshwater scampi production in the country.

ii. ***System diversification:*** Ponds and tanks form the basic resources for aquaculture in the country. However, there are huge other resources which have also been brought under the ambit of the culture system. Sewage fed fish farming is a typical example of harvesting wealth from waste in terms of producing fish protein. The huge low land rice field in the Brahmaputra valley have been effectively used for rice fish farming. This technology has also been used in other low land areas of the country. Integrated farming of compatible components including fish has served as an excellent method of nutrient and waste recycling for production of food including fish protein. Efforts have been made to refine the production process in all these above system orienting towards enhancing food production including fish protein. Flood plains have been effectively used either for seed rearing or short term fish farming.

High density fish farming has become a recent trend in the aquaculture sector. The popular ones are the use of the concrete tanks, biofloc system and recirculating aquaculture system (RAS). Since such systems offer fish farming in a more controlled environment, high density rearing is done to achieve greater fish yield. Biofloc systems can be extensively used for seed production with high seed yield and profit margin. Use of RAS have been started in some pockets for grow-out production of freshwater fishes. However, the production technologies in RAS are yet under the standardization process and expected to improve fish yield in the coming days.

iii. ***Change in cropping pattern:*** Optimized utilization of the pond production capacity largely involves judicious management of the gap between the standing crop and the carrying capacity. The carrying capacity of the pond is dynamic and depends on the inputs and adopted management protocols followed during the culture period. Though a larger gap between standing crop and carrying capacity and result in under-utilisation of productivity potential. Various cropping pattern have been devised for the grow-out ponds in recent years to ensure more effective fish production. Cropping methods such as single stock (more density)-multiple harvest (SSMH), multiple stock-multiple harvest (MSMH), multiple cropping and minor carp intercropping in cap ponds have been devised and evaluated with realisation of higher fish production than the conventional SSSH method. The biomass yield in intercrop, SSMH and MSMH have been demonstrated to yield 21, 17 and 24% higher than SSSH, and the respective net incomes were 32.9, 21.3 and 56.5% higher (Das *et al.*, 2019). All these cropping patterns not only shown better yield, but also increased the profit margin. Besides, the investment capacity of the farmers could be increased through scope of interim harvest, ploughing-back of investment and reducing the risk for the investment.

The grow-out farming practices have been evolved depending on the market forces and regional preference with change in species composition, stocking density and input use. The bi-species culture of rohu (80-90%) and catla (10-20%) with stocking of larger seed (100-200 g) has been evolved in the Koleru lake region of Andhra Pradesh. Whereas, the composite culture of the three or six major carp species is still in vogue in many States. Now-a-days, the commercial fish farming is gradually being transformed towards specialised and precision farming with a trend towards 'fish fattening process' where the stocking size ranges up to 400 g and crop duration is shortened to four to six months to produce the market size. This has also helped in creating a third tier of seed farmers, called stunted seed producer, after the fry and fingerling producer to cater the round the year seed demand.

iv. ***Breed improvement:*** Use of good quality seed with higher growth potential is an effective means to increase the pond productivity. Programme on stock improvement of important cultured species has long been started in the country. While selective breeding programmes for genetic improvement of rohu, catla, *M. rosenbergii* are ongoing at ICAR-CIFA, that of magur being carried out at ICAR-CIFE. The improved rohu released in 2005 as 'Jayanti rohu' has been showing at least 30-40% higher growth after the 12th generation in the farmers' fields and has become extremely popular among the farmers all over the country. Resistance against *Aeromonas* has also been added as a selection trait to further improve the 'Jayanti rohu'. Similarly, 30% higher growth has been demonstrated in farmers' fields for both 2nd generation catla and 10th generation in prawn. Improved magur also demonstrated higher growth in grow-out pond. Amur strain of common carp has been propagated in culture system with high degree of success in terms of its growth performance. The RGCA has been effectively popularising GIFT and red tilapia. While ICAR-CIFA is spreading the improved fish through 'Multiplier Centres' in several States, the National Freshwater Fish Brood Bank (NFFBB) established at Bhubaneswar under the administrative control of NFDB has been actively promoting the improved varieties. Further, State level Fish Brood Banks are established for wider dissemination of the improved cultivable species to the farmers' fields.

d) Use of non-conventional feed ingredients: Production and supply of formulated fish feed on commercial scale is one of the major interventions for development of freshwater aquaculture. Success of commercial production and supply of floating feed for striped catfish in the last decade has prompted use of such feed for other important freshwater species and the aquaculture sector has witnessed a shift towards feeding the fish with commercially available balanced diet. Large number of feed mills have been established in various parts of the country. These mills offer customised feed prepared from varied ingredients and as per the nutritional requirement of the cultured species and its life stages. Semi-sinking and floating feeds are commercially being supplied. Feed with balanced nutrition has helped in precise feed management and minimum loss to the environment and in turn, it helped in water quality maintenance in the pond ecosystem. ICAR-CIFA has developed specialized broodfish diet CIFABROOD™ which advances gonad growth and maturation, facilitates early spawning and significantly increases spawning response.

With increase in the area coverage and intensity of farming in the freshwater aquaculture sector, there has been a rise in the shortage of raw ingredients for feed production. While the aquaculture sector is already competing with dairy and poultry sector for these feed ingredients, a further rise in demand has made it imperative to search for alternate non-conventional ingredients for the fish feed preparation. Research efforts over the year have identified many non-conventional ingredients from available local resources. Farm-made feed are increasingly being used in many farms. Establishment of mini feed mill have been promoted through various sponsored schemes for small farms for their own consumption and supply to the local farmers. Many new plant products have been identified to substitute the fish protein in fish feed. In recent years, protein production of various insects such as the black soldier fly (BSF) have been explored and used effectively for commercial feed production. Feed fortified with nanoparticles for larval (CIFA Carp Starter) and grow-out stages (CIFA Carp Grower) for higher protein efficiency have been developed by ICAR-CIFA and commercialised.

e) ***Disease control and management:*** Rising intensity of farming is also associated with the increased occurrence of disease problems in the aquaculture sector. At the same time, it has also increased the awareness and attention of the farmers towards prevention and control of the disease outbreak and crop management. Health management of fish has gained importance in the sector following emergence of new disease. Diagnostic kits for early warning and detection of disease, chemical formulations and therapeutics for disease control are some of the latest supports made available for prevention and control of many common diseases. Development of CIFAX, the chemical formulation by ICAR-CIFA was a land mark achievement for effective control of EUS during 1990s. Few vaccines have been developed and many are under preparation against emerging diseases. Implementation of the National Surveillance Programme on Aquatic Animal Diseases (NSPAAD) is one of the significant development in the aquaculture sector in recent years providing active support to the aquaculture sector nationwide for surveillance and creation of database on disease emergence. Referral Laboratories for Level-III diagnosis have been established in the country to support an 'Emergency response system'. Such supports have increased the assurance of crop protection against disease and in turn has encouraged the investment in aquaculture operation.

f) ***Artificial intelligence and Machine learning:*** One of the key areas where data analytics has proven instrumental is in farm management systems. IoT devices and sensors deployed in aquaculture farms gather real-time data on critical parameters such as water quality, oxygen levels, and feed management. By analysing this data, farmers can gain valuable insights into the health and well-being of their animals and enable proactive measures to prevent disease outbreaks and optimize feed utilization. Furthermore, data analytics, coupled with artificial intelligence (AI) and machine learning (ML) have revolutionised aquaculture practices in India. AI-powered systems, coupled with extensive data analysis, enable farmers to make informed decisions and predict market trends. For instance, a study conducted by the Indian Council of Agricultural Research (ICAR) found that the implementation of AI algorithms in disease detection resulted in a 40% reduction in losses due to disease outbreaks.

Epilogue

The different approaches to innovation used in aquaculture development like seed production technology, effective technology dissemination for scientific fish farming, resource renovation and horizontal expansion, vertical increase in productivity, species diversification, system diversification, change in cropping pattern, breed improvement, use of non-conventional feed ingredients, disease control and management, artificial intelligence and Machine learning have been discussed in details. Aquaculture biotechnology and other technological innovations are showing a positive impact on aquaculture diversification success, investment potential, and international technology exchange. However, this development will largely depend on the desire and willingness of the producers to work hand to hand with scientists and the international donor community to assist developing counties in related research, capacity building and infrastructure development. Improved exchange of information and discussion between scientists, researchers, and producers from different regions on their problems and achievements will undoubtedly help this important sector to further develop with the view to increasing sustainable aquatic animal production globally.

Further Reading

Das P.C. and Ferosekhan S. 2022. Recent Advances in Carp Culture in India. In: Souvenir of IFO-2022 (Eds. Das B.K. *et al.*). ICAR-CIFRI, Barrackpore, ISBN: 8185482446

Das, P.C., 2018. Grow-out Fish Farming in Freshwater: Principles & Practices. Mass Breeding and Culture Technique of Catfishes, p.97

Das, P.C., 2019. Development of aquaculture for commercially important finfishes in India. Aquaculture of commercially important finfishes in South Asia. SAARC Agriculture Centre, SAARC, Dhaka, Bangladesh, pp.15-46

Das, P.C., Kamble, S.P., Velmurugan, P. and Pradhan, D., 2019. Evaluation of minor carps intercropping in carp polyculture vis-à-vis other grow-out cropping patterns of carp farming. Aquaculture Research, 50(6), pp.1574-1584

DoF, GoI, 2023. Annual Report 2022-23. Ministry of Fisheries, Animal Husbandry & Dairying, Govt. of India, New Delhi

FAO. 2022. The State of World Fisheries and Aquaculture 2022. Towards Blue Transformation. Rome, FAO. https://doi.org/10.4060/cc0461en

Jena, J. K., Gopalakrishnan, A., Ravisankar, C. N., Lal, K. K., Das, B. K., Das, P. C., *et al.* (2022). "Achievements in fisheries and aquaculture in independent India," in Indian Agriculture after Independence, eds H. Pathak, J. P. Mishra and T. Mohapatra (New Delhi: Indian Council of Agricultural Research), 426.

Jena, J.K. and Das, P.C., 2018. Assuring fish demand of India by 2030 through aquaculture development: Issues and Strategies. Indian Farming, 68(10).

Jena, J.K. and Das, P.C., 2022. New Paradigms in Freshwater Aquaculture in Coastal Ecosystems in India: Happiness and Hope. In Transforming Coastal Zone for Sustainable Food and Income Security: Proceedings of the International Symposium of ISCAR on Coastal Agriculture, March 16–19, 2021 (pp. 361-376). Cham: Springer International Publishing.

Swain S., Ferosekhan S. 2022. Present Status and Future Scope of Freshwater Aquaculture Sector in India. In: Souvenir of IFO-2022 (Eds. Das B.K. *et al.*). ICAR-CIFRI, Barrackpore, ISBN: 8185482446

8

Sustainability in Mariculture Through Innovative Farming Systems

Imelda Joseph and Shoji Joseph*

ICAR-Central Marine Fisheries Research Institute, Post Box No.1603, Ernakulam North P.O., Kochi, Kerala

**Email: imeldajoseph@gmail.com*

Introduction

The 2030 Agenda of UN General Assembly is encapsulating the 17 Sustainable Development Goals (SDGs), outlines a comprehensive framework for humanity's pursuit of a more prosperous, equitable, and sustainable future. Beyond addressing poverty, hunger, health, and nutrition, the agenda aspires to reduce inequalities and foster peaceful, just, and inclusive societies-all while respecting planetary boundaries. The Food and Agriculture Organization of the United Nations (FAO) projects a global population of 10 billion by 2050 and warns that existing food production systems and nutritional models are insufficient and unsustainable to meet this demand. With the projected global population, and a potential 88% increase in demand for animal proteins (Searchinger*et al.*, 2018), there is a pressing need to confront the challenge of providing a healthy and sustainable diet for a growing population, often exceeding recommended consumption levels. To address these challenges, efforts are underway to promote sustainable culture practices, explore alternative protein sources, improve aquaculture sustainability, reduce food waste, invest in genetic improvements, embrace technological innovations, and implement supportive policies. Collaborative actions across governments, organizations, and individuals are essential to build a resilient and sustainable global food system capable of feeding the growing population while minimizing environmental impact.

Fish contributes to over 30% of global protein consumption, with half of this supply originating from aquaculture and mariculture, as highlighted by Costello *et al.* (2020). In 2020, farmed fish production encompassed 57.5 million tons for finfish, 17.7 million tons for molluscs, 11.2 million tons for crustaceans,

5,25,000 tonnes of aquatic invertebrates, and 5,37,000 tonnes of semi-aquatic species including turtles and frogs. The rapid growth of aquaculture, outpacing other major food production sectors, is attributed to the low feed conversion ratio of cultured species (1.1–1.6 kg of feed per kg of edible fish), surpassing ratios for livestock like pork (FCR up to 4.4) or beef (FCR up to 9) (FAO, 2022). Despite this efficiency, aquaculture faces sustainability challenges due to the carnivorous nature of many fishes, necessitating diets rich in animal protein. To ensure future viability as a food source, the aquaculture industry must address food waste, promote sustainable practices, and encourage shifts toward healthier diets (Garcia-Oliveira *et al.*, 2020). This transformation is crucial to mitigate the environmental impact of the expanding aquaculture sector, projected to persist or even intensify globally (FAO, 2022).

Sustainable aquaculture, irrespective of its scale, is characterized by its economic viability and environmental responsibility. Beyond these criteria, particularly in regions where aquaculture and fisheries hold cultural significance, it must also align with cultural values and not compromise access to vital resources for small-scale fishers and local communities. Examples of environmentally sustainable aquaculture practices include integrated multi-trophic aquaculture, seaweed aquaculture, shellfish aquaculture, and thoughtfully planned fish rearing employing an ecosystem-based approach. These methods emphasize minimizing environmental impact, optimizing resource utilization, and fostering harmony with the cultural and social fabric of the communities involved. In essence, the pursuit of sustainability in aquaculture involves a holistic approach that balances economic, environmental, and socio-cultural considerations for the long-term benefit of both ecosystems and communities.

The diverse nature of the aquaculture sector, spanning various species, production systems, and scales, holds the potential for multifaceted contributions to the Sustainable Development Goals (SDGs). Through the culture of different aquatic species and utilization of varied production systems, aquaculture can offer a range of nutritional, economic, and environmental benefits. Methodologies such as Life Cycle Assessments, economic evaluations, and social impact assessments help identify and measure these contributions, ensuring a comprehensive understanding of aquaculture's role in achieving SDGs. However, trade-offs exist, requiring a delicate balance between economic gains, environmental sustainability, and social equity. Effective policies and strategies are crucial to harness the sector's diversity for positive contributions to SDGs, addressing challenges while maximizing its potential to support a sustainable and equitable future.

Sustainable Aquaculture

Sustainability is the capacity of human activities to persist in time while maintaining a healthy environment. It is further defined as "Use of the environment and resources that meets the needs of the present without compromising the ability of future generations to meet their own needs" (Brundtland definition). Sustainable biological systems are characterized as diverse, adaptable, resilient, and productive over time. Sustainable aquaculture entails social, economic and environmental sustainability.

Sustainable farming systems

Aquaculture plays a pivotal role in meeting the global demand for animal protein, but ensuring its long-term viability necessitates a transition to sustainability.

Cage fish farming

Cage fish farming is a method of rearing fish in floating or submerged cages within natural water bodies, such as lakes, rivers, or in sea. This farming technique is widely employed for various species, including salmon, tilapia, and sea bass. The cages are typically constructed with netting or mesh materials, allowing water to flow freely through them while confining the fish within a designated area. Cage fish farming offers several advantages, including efficient space utilization, reduced environmental impact, ease of management and operation, and the potential for high stocking densities and therefore high production. However, challenges associated with cage farming include concerns about water quality, disease management, and the escape of farmed fish, which can impact wild populations. The industry continually explores technological innovations and best management practices to address these challenges and enhance the sustainability of cage fish farming, contributing to the global supply of seafood.

Advantages of cage fish culture

- Productive utilization of water resources
- Diverse species for farming
- Low initial investment is required in an existing body of water
- Harvesting is simple
- Monitoring of fish is simple
- Management is easier
- Less labour intensive
- Product quality is assured
- Alternate employment and income

Environmental concerns in cage farms

Outputs from cage farms released to the ambient water and,

- Cause eutrophication
- Spread disease in the wild
- Increase BOD, COD
- Increase TAN, TN, TP levels
- Increase sedimentation (TSS), and
- Emission of GHGs

Cage fish farming operations can enhance their overall sustainability, balancing economic viability with environmental and social responsibility by focusing on several key aspects as given below:

Quality Seed: Utilizing high-quality seed, preferably produced in hatcheries, is essential for ensuring the health, disease resistance, and overall robustness of the fish stock. Selecting species that are well-suited to the local environment and have desirable market value is crucial for the economic sustainability of the operation.

Good Water Quality: Maintaining optimal water quality is fundamental for the well-being of farmed fish. This involves minimizing waste output, ensuring good water flow within the cages, and preventing pollution. Regular monitoring and management of water parameters contribute to a healthy aquatic environment, reducing the risk of disease outbreaks and promoting sustainable production.

High-Quality Feed: Providing nutritionally complete and cost-effective feed is a key factor in sustainable cage fish farming. Focus on achieving a low Feed Conversion Ratio (FCR), which indicates efficient conversion of feed into fish biomass, and minimizing feed wastage. Additionally, exploring alternatives and substitutions for fish meal and fish oil in the feed, such as plant-based proteins and sustainable sources, supports environmental sustainability and reduces reliance on marine resources.

Fish Meal/Fish Oil Substitution: In the context of achieving a high-protein diet for farmed fish, the sustainable substitution of fish meal and fish oil is crucial. Identifying alternative protein and lipid sources that are both nutritionally adequate and environmentally friendly helps reduce the reliance on wild-caught fish for feed, contributing to the overall sustainability of the aquaculture operation.

Ensuring the sustainability of cage fish farming involves a series of strategic practices and considerations.

a) ***Appropriate Site Selection***: Carefully choose cage installation sites based on environmental conditions and suitability to minimize negative impacts on surrounding ecosystems.

b) ***Adherence to Carrying Capacity***: Install cages in accordance with the carrying capacity of the water resources, preventing overloading and maintaining ecological balance.

c) ***Optimal Stocking Levels***: Stock cages with fish at levels that align with the carrying capacity of the system, ensuring optimum production without exceeding environmental limits.

d) ***Nutritionally Complete Feed***: Provide fish with nutritionally complete and quality feed to minimize waste and control the release of unutilized nutrients, promoting both economic efficiency and environmental sustainability.

e) ***Avoidance of Antibiotics and Drugs***: Minimize the use of antibiotics and drugs to promote fish health, and explore alternative disease prevention and control methods to reduce environmental impact.

f) ***Multi-Species Stocking***: Introduce compatible multi-species in cages, fostering ecological balance and reducing the risk of disease and pest outbreaks.

g) ***Escape Prevention***: Implement measures to prevent fish escape from cage farms, avoiding potential negative impacts on local ecosystems and biodiversity.

h) ***Preference for Local Species***: Prioritize farming species of local importance to prevent adverse effects on biodiversity in case of escape and promote the sustainability of local ecosystems.

i) ***Avoidance of Genetically Manipulated and Exotic Species***: Refrain from farming genetically manipulated species and exotic species in cages to prevent unintended consequences on local biodiversity and ecosystems.

j) ***Protection of Natural Fish Habitats***: Avoid cage installation in areas with natural fish habitats to protect and preserve the diversity of these environments.

k) ***Promotion of Product Certification***: Encourage and participate in product certification programs to inform consumers about the sustainability practices employed in cage fish farming operations.

By implementing these practices, cage fish farmers can contribute to the long-term sustainability of their operations, minimizing environmental impact, promoting ecosystem health, and meeting the increasing demand for responsibly sourced food.

Integrated multi-trophic aquaculture (IMTA)

Integrated Multi-Trophic Aquaculture (IMTA) stands out as a pioneering and sustainable approach to aquaculture, emulating the intricacies of natural ecosystems by cultivating diverse species from varying trophic levels within a shared aquatic environment. At its core, IMTA orchestrates symbiotic relationships among species, strategically combining carnivorous fish, primary-producing seaweeds, and filter-feeding bivalves. The system's ingenuity lies in harnessing nutrient cycling, where waste from one species becomes nourishment for another. Fish waste, for example, becomes nutrients for seaweed and bivalves, fostering a balanced and eco-friendly environment. IMTA not only minimizes the ecological footprint of aquaculture, enhancing environmental sustainability, but also optimizes resource usage, boosting economic efficiency.

In IMTA, fed species coexist with extractive species that capitalize on inorganic and organic wastes, transforming them into valuable resources like fertilizer, food, and energy for other crops. This cyclical approach not only recaptures otherwise lost nutrients but also facilitates bio mitigation by removing carbon dioxide and partially extracting nutrients, simultaneously producing oxygen. The extractive component of IMTA serves a dual purpose, generating versatile biomass and rendering waste reduction services. Importantly, every farmed component within IMTA holds commercial value, contributing to both economic viability and environmental stewardship. The intricate mix of organisms across different trophic levels in IMTA mirrors the resilience and functionality of natural ecosystems, presenting a transformative model for sustainable and responsible aquaculture practices.

IMTA systems can be implemented in various aquatic environments, including coastal areas and open water, and they are regarded as a promising solution to address some of the environmental challenges associated with traditional monoculture practices. As the aquaculture industry continues to seek more sustainable practices, IMTA represents a forward-looking approach that aligns with principles of ecological balance and resource efficiency. IMTA is a solution to eutrophication caused by feed aquaculture by extraction and conversion of the excess nutrients and energy into other commercial crops. It is an innovative solution promoted for environmental sustainability, economic stability and societal acceptability.

The by-products and wastes generated by aquaculture operations can be repurposed through circular economy principles, contributing to sustainability and economic throughput. These waste streams can find new life as ingredients in food, cosmetics, or pharmaceuticals, aligning with the circular concept of converting waste into valuable resources. By adopting such circular practices, the aquaculture industry can not only meet protein demands but also contribute to a more sustainable and economically efficient model for the future.

Offshore Cage farming

Offshore cage farming aims to minimize environmental impacts on coastal ecosystems, reduce interactions with wild marine species, and alleviate concerns related to coastal water quality.

Key Considerations for Offshore Mariculture Development

Disease and Parasite Mitigation: Offshore locations have the potential to reduce the risk of diseases and parasites affecting farmed species. The greater water depth and distance from coastal areas can minimize the transfer of pathogens from wild to farmed populations.

Algal Bloom Avoidance: Offshore locations may experience fewer issues related to algal blooms, which can be a concern in inshore areas. The increased water exchange and distance from nutrient-rich coastal zones can contribute to a more stable and controlled environment.

User Conflicts: Moving mariculture offshore can mitigate conflicts with other coastal users, such as recreational activities or shipping lanes. This may lead to improved social acceptance and reduced regulatory challenges.

Seaweed and Bivalve Profitability: Seaweed and bivalve cultivation can be particularly suitable for offshore development. These organisms do not require feeding, relying on nutrients available in the water column, and operational costs can be lower compared to fed species.

Challenges and Considerations

Infrastructure and Technology: Developing infrastructure and technology capable of withstanding the challenges of offshore conditions, including strong currents and waves, is crucial for the success of offshore mariculture.

Monitoring and Management: Remote offshore locations pose challenges for monitoring and managing mariculture operations. Advanced technologies for real-time monitoring and adaptive management strategies are essential.

Regulatory Frameworks: The development of appropriate regulatory frameworks is critical to govern offshore mariculture and address environmental, economic, and social considerations.

Environmental Impact: While offshore mariculture can potentially reduce impacts on inshore ecosystems, it is essential to carefully assess and mitigate any potential environmental consequences, such as nutrient discharge and interactions with open ocean biodiversity.The shift toward offshore mariculture development aligns with the goal of expanding the aquaculture industry while addressing environmental and sustainability concerns. Careful planning, technological innovation, and collaboration among stakeholders are essential to navigate the complexities of offshore aquaculture and ensure its long-term success.

Precision Fish Farming (PFF)

Precision Fish Farming (PFF) represents a recent and innovative framework in fish farming that has evolved from the concept of Precision Livestock Farming (PLF). This approach, as outlined by Norton and Berckmans in 2018, leverages advanced hardware such as sensors, observers, and intelligent software to enhance various aspects of fish culture. The primary goals of Precision Fish Farming include improving animal health and welfare, increasing overall productivity, enhancing yield, and promoting environmental sustainability within aquaculture operations.

Key Components and Objectives of Precision Fish Farming (PFF)

Sensor Technology: PFF integrates sensor technologies that monitor and collect data on various parameters such as water quality, feeding behavior, and fish health. These sensors provide real-time information, allowing for prompt and informed decision-making.

Intelligent Software: Advanced software solutions analyze the data collected by sensors, enabling farmers to make precise adjustments in fish farming practices. This includes optimizing feeding regimes, health management strategies, and environmental conditions within the farming system.

Animal Health and Welfare: PFF emphasizes the improvement of animal health and welfare through proactive monitoring. Early detection of diseases, anomalies, or stressors allows for timely interventions, reducing the need for antibiotics and mitigating health challenges.

Productivity and Yield Optimization: By fine-tuning various parameters based on real-time data, PFF aims to enhance overall productivity and yield. This includes optimizing feeding strategies, growth rates, and resource utilization, leading to more efficient and sustainable fish farming operations.

Environmental Sustainability: PFF contributes to environmental sustainability by minimizing resource wastage and reducing the environmental impact of fish farming. Precise control over feeding, water quality, and waste management helps minimize the ecological footprint of aquaculture operations.

Challenges and Considerations in PFF

Technological Adoption: The successful implementation of PFF relies on the widespread adoption of advanced technologies, which may pose challenges for some farmers in terms of investment, training, and infrastructure.

Data Security: Handling and managing the vast amount of data generated by PFF systems raise concerns related to data security and privacy. Robust measures must be in place to safeguard sensitive information.

Integration with Traditional Knowledge: Integrating PFF with traditional knowledge and practices in fish farming is crucial to ensure that technological advancements complement existing expertise and contribute to sustainable and culturally appropriate aquaculture practices. Precision Fish Farming holds promise as a forward-thinking and sustainable framework, offering the aquaculture industry opportunities to enhance efficiency, reduce environmental impact, and prioritize animal welfare through the integration of cutting-edge technologies and data-driven decision-making.

Circular aquaculture

Circular aquaculture represents a transformative approach to fish farming, embodying principles of sustainability and waste reduction. Unlike traditional linear production systems, circular aquaculture aims to create closed-loop cycles, repurposing waste products as valuable inputs. This innovative model integrates multiple species and components, such as algae or bivalves, to create balanced ecosystems within aquaculture operations. By emphasizing water recycling, minimizing environmental impact, and promoting responsible resource use, circular aquaculture addresses challenges associated with conventional fish farming. This shift towards a holistic and regenerative model aligns with environmental and conservation goals, highlighting the industry's dedication to sustainable practices for long-term viability. The application of circular economy principles in aquaculture further enhances sustainability by optimizing resource use, transforming waste into valuable products like food, feed, bio-based materials, and bioenergy. Circular business models offer the potential to significantly reduce waste, enhance efficiency, and contribute to a more sustainable and resource-efficient future for aquaculture. It is imperative for the industry to lead in adopting circular approaches, guided by regulators and informed by research and innovation, with indicators and reporting methods playing a crucial role in mainstreaming circularity practices into industry norms and policies.

a) **Innovative application of aquaculture sub-products:** Even with the application of the Integrated Multi-Trophic Aquaculture (IMTA) model, aquaculture generates various sub-products. In circular aquaculture, these by-products are systematically repurposed to minimize their quantity and prevent environmental release, aligning with the industry's commitment to sustainable and responsible waste management. Sub-products resulting from the processing of primary fish products, such as fish bone, skin, belly flaps, or trimmings, are recognized for their significant potential. These by-products hold value as they can be reintegrated into the food chain as ingredients for human consumption, maximizing their benefit and reducing waste. The utilization of these sub-products underscores a sustainable approach within the seafood industry, contributing to a more efficient and circular use of resources. Another category of wastes in aquaculture encompasses diverse materials, including manure, non-mineralized guano, and digestive tract content, along with animal by-products containing authorized residues surpassing permitted levels. This category also includes deceased animals, oocytes, and embryos, carcasses, as well as parts of animals slaughtered that are not intended for human or animal consumption due to legal or commercial reasons. Managing these waste categories is essential for maintaining environmental sustainability and regulatory compliance within the aquaculture industry.

b) **Human food ingredients:** Aquaculture sub-products offer a sustainable reservoir for reclaiming unconsumed compounds like proteins, lipids, and pigments, predominantly repurposed for the production of broths, aromas, fish protein concentrates, among other products. Within the food industry, various compounds crucial for human consumption can be derived from these sub-products, including fish flour, chitosan (a biopolymer sourced from chitin in crustacean exoskeletons), concentrated proteins, collagen (primarily extracted from fish skins), and gelatin (produced through partial collagen hydrolysis). Astaxanthin and chitosan, obtained from aquaculture sub-products, are not only vital in industrial applications but also serve as nutritional supplements for humans. Astaxanthin, a prevalent carotenoid obtained from sub-products, particularly in salmon, trout, krill, shrimps, freshwater crabs, and crustacean shells, is widely recognized for its application as a food ingredient due to its colour-enhancing and antioxidant properties. The cultivation of macro and microalgae has become crucial in this domain, offering a diverse source of pigments that can be recovered from sub-products after processing. Beta-carotene, a yellow pigment known as a precursor to vitamin A, is among the pigments that can be obtained, providing both colour and additional properties, such as antioxidant benefits, to various products destined for human consumption.

c) **Animal feeding:** Various sub-products stemming from aquaculture, including fish flour, ground shell, chitosan, astaxanthin, concentrated proteins, and silage resulting from fish liquefaction, present valuable opportunities for incorporation into feed formulations. These diverse components can be utilized in the formulation of feeds for aquaculture animals, farm animals, and pets. By integrating these aquaculture-derived sub-products into feed formulations, the industry can enhance the nutritional content of feeds while also contributing to a more sustainable and circular use of resources. This practice aligns with the broader goal of optimizing resource efficiency within the agricultural and aquaculture sectors. Crushed shells, particularly from oysters, serve as a significant source of calcium supplementation ($CaCO_3$), proving highly beneficial when incorporated into hen feeding. Substituting the calcium derived from limestone with oyster shells has been demonstrated to improve egg production, enhance the strength, weight, and thickness of eggs. Additionally, chickens fed with oyster shells exhibited a more rapid increase in weight. This practice underscores the value of utilizing natural and sustainable sources, such as oyster shells, to enhance the nutritional content of poultry feed and promote overall animal health and productivity.

d) **Agriculture:** Mussel shells serve a valuable role as a liming agent and soil amendment in farming practices. Comprised mainly of calcium carbonate ($CaCO_3$), these mussel shells offer a natural resource that effectively neutralizes acidic and metal-contaminated soils. Simultaneously, they contribute to soil fertility and elevate oxygen levels. The utilization of mussel shells as a natural product aligns with ecological agriculture principles, providing a sustainable alternative to mined-$CaCO_3$. This practice showcases the potential of repurposing marine by-products for soil improvement, demonstrating a more environmentally friendly approach in agriculture.

e) **Industrial uses: food packaging, cosmetic and pharmaceutical:** Marine protein-based products, including collagen and gelatine, along with lipids and pigments, have proven to be valuable resources for industries such as food packaging, cosmetics, and pharmaceuticals. These marine-derived compounds offer diverse applications due to their unique properties. Polyunsaturated fatty acids (PUFAs) from marine sources have been subjects of extensive research, showcasing their potential applications, particularly in the realms of nutraceutical and medicine. The exploration of marine biomolecules underscores the rich potential of the oceans as a source of innovative and sustainable ingredients for various industrial sectors.

The food industry is adopting biodegradable active packaging to reduce single-use plastics and enhance product shelf-life. Chitosan, gelatine, and carrageenan are commonly used for biodegradable packaging, while active ingredients from algae extracts, squid, or *Litopenaeus vannamei* by-products are included to improve preservation properties. This sustainable approach aligns with efforts to minimize environmental impact and promote innovative packaging solutions. Pigments serve as colour sources with added bioactive properties. Astaxanthin, known for its antioxidant properties, is reported to stimulate the immune system and prevent various diseases, including diabetes, cardiovascular issues, and neurodegenerative conditions. In cosmetics, astaxanthin is utilized in skincare and anti-aging formulations. Beta-carotene, precursor of vitamin A, exhibits antioxidant activity that counteracts the impact of free radicals, potentially preventing oxidative damage associated with inflammatory processes and chronic diseases like diabetes, cardiovascular issues, and cancer. Pigments from microalgae, suitable for human or animal consumption, offer good digestibility due to a simpler matrix than higher plants, facilitating their incorporation as nutritional ingredients.

The emergence of circular bio-economy aquaculture production systems signifies an innovative approach that not only fosters sustainability but also has the potential to generate new job opportunities. By following circular bio-economy principles, aquaculture can contribute significantly to economic growth while prioritizing resource efficiency and environmental stewardship. This integrated approach aligns with the broader goal of building resilient and sustainable aquaculture production systems that benefit both local communities and the environment.

References

Costello, C., Cao, L., Gelcich, S. *et al.* The future of food from the sea. Nature 588, 95–100 (2020). https://doi.org/10.1038/s41586-020

FAO.2022. The State of World Fisheries and Aquaculture 2022. Towards Blue Transformation. Rome, FAO. https://doi.org/10.4060/cc0461en.

Fraga-Corral, M., Rona, P., Garcia-Oliveira, P., Pereira, A.G., Losada, A.P., Prieto, M.A., Quiroga, M.I., and Simal-Gandara, J. 2022. Aquaculture as a circular bio-economy model with Galicia as a study case: How to transform waste into revalorized by-products, Trends in Food Science & Technology, 119: 23-35.

Garcia-Oliveira, P., Fraga-Corral, M., Carpena, M., Prieto, M.A., and Simal-Gandara, J. 2022. Approaches for sustainable food production and consumption systems. In: Future Foods Global Trends, Opportunities, and Sustainability Challenges, Academic Press. 23-38pp.

Searchinger, T. D, Wirsenius, S, Beringer, T., and Dumas P. 2018. Assessing the efficiency of changes in land use for mitigating climate change. Nature, 564: 249–253.

9

Fish Harvest Technology Status and Way Forward

Pravin Putra

Former Assistant Director General (Marine Fisheries), Indian Council of Agricultural Research, New Delhi

Email: pravinp2005@gmail.com

Introduction

Fish harvest technology has evolved significantly over the years, from traditional methods to modern techniques. These advancements have had a significant impact on the fishing industry, fish populations, and marine ecosystems. Traditional fish harvest methods have been used for centuries. However, modern fish harvest technology, such as trawling, gill netting, long lining and purse seining, have been introduced to increase the efficiency of fish harvesting. These modern techniques have advantages, such as higher catch rates and reduced labour costs, but they also have disadvantages. For example, trawling can cause damage to the seafloor and non-target species, while purse seining, long lining and gill netting can result in by-catch of unwanted species.

The introduction of modern fish harvest technology has led to changes in fishing methods and practices. Fishers now have access to larger and more efficient vessels, which has led to overfishing in some areas. This has had a significant impact on fish populations and marine ecosystems, leading to declines in some species and changes in the food web. Furthermore, the use of modern fish harvest technology has also impacted fishers' livelihoods and income, as traditional methods are often unable to compete with the efficiency of modern techniques.

Fish harvesting technology plays a crucial role in the fishing industry and the broader context of food security, economic development, and environmental sustainability. The importance of fish harvest technology lies in its role in ensuring the sustainable management of fish stocks and the long-term viability of fisheries. By utilizing both traditional and modern technologies, fish harvesters can effectively and efficiently capture fish while minimizing negative impacts on the marine ecosystem. Additionally, technology plays a crucial role in stock assessment and monitoring, providing valuable data for informed decision-making in fisheries management. Furthermore, technological innovations, such as precision fishing, AI, drones, and remote sensing, contribute to more selective and sustainable harvesting, addressing challenges such as overfishing, by-catch, habitat destruction, and climate change.

Importance of fish harvesting technology

Efficiency and productivity: Advanced fish harvesting technology, such as modern fishing vessels, fishing gears, and processing equipment, can significantly increase the efficiency and productivity of fishing operations. This allows for larger quantities of fish to be harvested in a shorter amount of time, meeting the demand for seafood products while minimizing the impact on fish populations.

Sustainable fisheries management: Technological advancements in fish harvesting can contribute to sustainable fisheries management by enabling more precise and selective fishing methods. This can help reduce by-catch (the unintentional capture of non-target species), minimize habitat damage, and aid in the conservation of fish stocks.

Economic impact: Fish harvesting technology supports the economic development of coastal communities and fishing industries by creating employment opportunities, increasing the value of fishery products through improved handling and processing, and enhancing the overall competitiveness of the sector.

Food security and nutrition: Fish is a vital source of protein and essential nutrients for billions of people around the world. By improving the efficiency and effectiveness of fish harvesting, technology helps ensure a more reliable supply of fish for human consumption, contributing to global food security and nutrition.

Safety and working conditions: Modern fish harvesting technology can enhance the safety and working conditions of fishermen by reducing physical labour, minimizing exposure to hazardous conditions, and improving overall operational safety.

Innovation and research: The development of fish harvesting technology drives innovation and research in areas such as marine engineering, materials science, and robotics, leading to new solutions for sustainable fishing practices and the responsible utilization of aquatic resources. Fish harvesting technology is essential for the sustainable management of fisheries, economic development, food security, and the well-being of coastal communities. Advancements in this field can help address the challenges facing the fishing industry while promoting the responsible and efficient utilization of marine resources

Brief overview of traditional and modern fishing methods

Traditional Fishing Methods: (a) Hand Gathering: This involves catching fish by hand, using tools such as spears, nets, or bare hands. (b) Traps and Weirs: Fish traps and weirs are constructed using natural materials or simple tools to catch fish as they swim by. (c) Harpoons: Harpoons are thrown or thrust into the water to impale fish, particularly in shallow waters or from small boats. (d) Angling: Angling involves using a fishing rod, line, and hook to catch fish. This method has been practiced for centuries and is still widely used today. (e) Drag nets: used in near shore waters and operated manually by group of fishermen.

Modern Fishing Methods: (a) Trawling: Trawling involves dragging a large net through the water to catch fish. This method is commonly used in commercial fishing. (b) Gillnetting - fishing method that uses gillnets and the fish are caught by gilling. Gillnets can be characterized by mesh size, as well as colour and type of filament from which they are made. This is a highly selective fishing gear. (c) Long-lining: Long-lining involves setting a long line with baited hooks to catch fish such as tuna and swordfish. (d) Purse Seining: This method involves surrounding a school of fish with a large net and then drawing the bottom of the net closed to capture the fish.

Current status of fish harvest technology

In the modern era, fish harvest technology has seen significant advancements in various aspects of the fishing process, including locating, catching, and processing fish. One of the most notable developments is the use of sonar and satellite technology for fish detection and tracking. These technologies enable fishers to identify fish schools with greater accuracy, leading to more efficient harvesting practices.

Furthermore, the use of advanced fishing gear such as trawls, purse seines, and longlines has increased the industry's capacity to catch larger quantities of fish. Additionally, innovations in on-board processing and preservation techniques have extended the shelf life of harvested fish, allowing for longer transport and storage times. The technology used in fish harvesting has evolved over time, and the current status of fish harvest technology includes a wide range of methods and equipment to capture and process fish.

Many traditional methods of fish harvesting, such as net fishing, line fishing, and trapping, are still widely used today, especially in small-scale and artisanal fisheries. These methods have been refined over generations and continue to be an important part of the Indian fishing industry.

The fishing industry has seen significant advancements in gear technology, including the development of more efficient and selective fishing gear to reduce by-catch and minimize environmental impact. Innovations such as modified trawl designs, eco-friendly nets, and more selective fishing gear have been developed to improve the sustainability of fishing operations. The use of automation and robotics in fish harvesting is an area with significant potential for advancement. Automated systems for tasks such as fish sorting, processing, and packaging can improve efficiency and reduce labour costs. Additionally, autonomous or remotely operated fishing vessels are being developed to improve the safety and efficiency of fishing operations.

The integration of data and monitoring technologies, such as GPS, sonar, and satellite imagery, has enabled fishers to make more informed decisions about where and when to fish. These technologies can help in locating fish stocks, avoiding overfished areas, and complying with regulations.

To ensure the sustainability of fish harvest technology, it is important to promote sustainable harvest practices. This includes reducing overfishing, minimizing by-catch, and protecting marine ecosystems. Technological innovations, such as the use of sonar and satellite tracking, can help fishers locate fish more efficiently and reduce the impact on the environment. Additionally, policies and regulations can be implemented to promote sustainable fishing practices, such as limiting fishing quotas and protecting critical habitats.

Overfishing and by-catch

Overfishing and by-catch are two significant issues facing the world's fisheries, and they have serious implications for marine ecosystems and the long-term sustainability of fish stocks.

Overfishing: It occurs when fish are caught at a rate that exceeds the ability of the species to replenish itself through natural reproduction. This can lead to the depletion of fish populations, disruption of marine food webs, and the collapse of fisheries. Overfishing is often driven by a combination of factors, including technological advances in fishing methods, increased demand for seafood, and inadequate fisheries management.

By-catch refers to the unintentional capture of non-target species in fishing gear. This can include species such as seabirds, sea turtles, marine mammals, and non-commercial fish species. By-catch can result in significant ecological impacts, as well as economic and social consequences for fishing communities. It is often a result of using non-selective fishing gear or fishing in areas with high biodiversity.

Efforts to address overfishing and by-catch

Sustainable fisheries management: Implementing science-based fisheries management practices, including setting catch limits, establishing marine protected areas, and implementing seasonal closures, to allow fish populations to recover and maintain healthy levels.

Improved gear technology: Developing and implementing more selective fishing gear, such as escape panels, turtle excluder devices, and modified trawl designs, to reduce by-catch and minimize the impact on non-target species.

Monitoring and enforcement: Strengthening monitoring and enforcement of fishing activities to ensure compliance with regulations, prevent illegal, unreported, and unregulated (IUU) fishing, and reduce the risk of overfishing and by-catch.

Market-based initiatives: Encouraging consumer demand for sustainably sourced seafood through certification programs, eco-labelling, and sustainable seafood guides, which can incentivize responsible fishing practices and support fisheries that prioritize sustainability.

International cooperation: Promoting international cooperation & agreements to address trans-boundary fisheries management and the conservation of shared fish stocks, as well as to combat illegal fishing activities.

Addressing overfishing and by-catch requires a multi-faceted approach that involves collaboration among governments, fishing industries, conservation organizations, and consumers. By implementing sustainable fishing practices and supporting responsible fisheries management, it is possible to mitigate the negative impacts of overfishing and by-catch and work towards the long-term health of marine ecosystems and fisheries.

Environmental Impacts

The development and use of fish harvest technology can have various environmental impacts, both positive and negative. It's important to consider these impacts when evaluating the sustainability and overall effects of modern fishing practices. Here are some environmental impacts associated with fish harvest technology:

Positive Environmental Impacts

Selective Harvesting: Advanced fishing technologies, such as selective gear and practices, can help reduce by-catch and minimize the unintended capture of non-target species. This supports biodiversity and reduces the impact on vulnerable species.

Reduced Fuel Consumption: Advanced navigation technology, such as GPS and sonar, can help fishermen locate fish more efficiently, and reducing the need for extensive searching and minimizing fuel consumption and greenhouse gas emissions. Improved designs of fishing crafts and gears also substantially save energy and reduce fuel consumption.

Habitat Protection: Some modern fishing technologies and practices aim to minimize damage to sensitive marine habitats, such as coral reefs and seagrass beds, helping to preserve these critical ecosystems.

Negative Environmental Impacts

Overfishing: The use of advanced fishing technologies, such as large-scale trawlers and longline fishing vessels, can contribute to overfishing if not managed sustainably. Overfishing can lead to the depletion of fish stocks and disrupt marine ecosystems.

Habitat Destruction: Certain fishing methods, such as bottom trawling, can cause significant damage to seafloor habitats and result in the destruction of marine ecosystems, including coral reefs and benthic communities.

By-catch and Discards: Despite advances in selective harvesting technology, some fishing practices still result in significant by-catch and discards, which can have negative impacts on non-target species and marine food webs.

Pollution: Fishing operations can contribute to marine pollution through the release of plastics, fuel spills, and waste materials, as well as due tothe use of antifouling paints on vessels. Furthermore, the reliance on fossil fuel-powered vessels has contributed to carbon emissions and ocean pollution, posing a threat to marine ecosystems. The need for sustainable fishing practices and environmentally friendly technologies is paramount to address these challenges and ensure the long-term viability of the fishing industry.

To mitigate the negative environmental impacts of fish harvest technology, sustainable fishing practices, ecosystem-based management, and the adoption of technologies that minimize environmental harm are crucial. Additionally, international cooperation, strong regulations, and responsible industry practices can play important roles in promoting environmentally sustainable fish harvesting.

Technological limitations and accessibility in fishing technology can pose significant challenges, particularly for small-scale and artisanal fishing communities. These limitations can impact the ability of these communities to compete effectively, ensure sustainability, and adapt to changing environmental conditions. Here are some key considerations related to technological limitations and accessibility in fishing technology:

Technological limitations and accessibility

Cost: Advanced fishing technologies, such as modern vessels, sonar equipment, and automated systems, can be prohibitively expensive for many small-scale fishing operations. The initial investment, as well as ongoing maintenance and training costs, can create financial barriers.

Training and Skills: Many advanced fishing technologies require specialized training and skills to operate effectively. Access to training programs, education, and technical support may be limited in some regions, particularly for small-scale fishermen in remote areas.

Information and Data Accessibility: Access to real-time data, such as weather forecasts, market prices, and fish stock assessments, can be limited in some fishing communities, impacting the ability to make informed decisions and optimize fishing operations.

Infrastructure: In some regions, the lack of adequate infrastructure, such as ports, cold storage facilities, and processing centres, can hinder the adoption of modern fishing technologies and the efficient handling of catch.

Regulatory Compliance: Compliance with fisheries regulations and standards, including those related to gear technology, and environmental impact, can be challenging for small-scale fishermen who may lack resources and capacity to meet these requirements.

To address these challenges, efforts can be made to promote the development and adoption of fishing technologies that are appropriate, affordable, and sustainable for small-scale and artisanal fishing communities.

The integration of data-driven approaches in fisheries management and conservation

The integration of data-driven approaches in fisheries management and conservation is becoming increasingly important for making informed decisions, improving sustainability, and reducing the impact of fishing activities on marine ecosystems. Data-driven approaches leverage various sources of data, including satellite imagery, electronic monitoring, acoustic sampling, and catch data, to inform management practices and support evidence-based decision-making. Here are some key aspects of the integration of data-driven approaches in fisheries:

Monitoring and Surveillance: Data-driven technologies, such as electronic monitoring systems, satellite tracking, and vessel monitoring systems (VMS), enable real-time monitoring of fishing activities, allowing authorities to track vessel movements, verify compliance with regulations, and identify potential illegal fishing activities.

Ecosystem-Based Management: Data-driven approaches support ecosystem-based management by integrating biological, ecological, and environmental data to assess the broader impacts of fishing on marine ecosystems. This holistic approach helps account for the interactions between target species, non-target species, and habitat characteristics.

Spatial Analysis and Planning: Geographic information systems (GIS) and spatial analysis techniques are used to map fishing activities, identify high-biodiversity areas, and designate marine protected areas. These tools help optimize spatial planning to minimize conflicts between conservation goals and fishing activities.

Data-driven technologies, such as electronic monitoring and underwater cameras, aid in quantifying and understanding by-catch, allowing for the development and testing of innovative gear modifications and fishing practices to reduce unintended capture of non-target species.

Research and Development

Investing in research and development to create technologies that are specifically tailored to the needs and limitations of small-scale fishing operations, including low-cost gear, energy-efficient vessels, and sustainable aquaculture systems. By addressing technological limitations and enhancing accessibility to modern fishing technologies, it is possible to support the livelihoods of small-scale fishermen, promote sustainability, and contribute to the long-term health of marine ecosystems.

Several technological innovations have been developed to help reduce by-catch in commercial fishing operations. These innovations aim to minimize the unintentional capture of non-target species, thereby supporting sustainable fishing practices and the conservation of marine biodiversity. Here are some examples of technological innovations for reducing by-catch:

Selective Fishing Gear: Innovative fishing gear designs, such as modified trawl nets, escape panels, and sorting grids, have been developed to allow the escape of non-target species while retaining the target catch. These selective gear technologies help reduce by-catch and minimize the impact on vulnerable species.

By-catch Reduction Devices: Various specialized devices, such as excluder grids for trawl fisheries and escape catches for crab and lobster pots, have been developed to allow non-target species to escape from fishing gear, thereby reducing by-catch.

Turtle Excluder Devices (TEDs): These are specialized devices installed in shrimp trawl nets to allow sea turtles and other large animals to escape from the net. These devices have been effective in reducing sea turtle by-catch in shrimp fisheries.

By-catch Reduction in Long Line Fisheries: Innovations such as circle hooks, which are less likely to catch non-target species, and bait modification techniques have been developed to reduce by-catch in long line fisheries.

Bird Deterrent Devices: Bird deterrent devices, such as streamer lines, underwater bait setting, and bird-scaring lines, are used to deter seabirds from diving into fishing gear, reducing the risk of by-catch in long line and trawl fisheries.

Acoustic Deterrents: Acoustic devices, including pingers and seal scarers, emit sounds that deter marine mammals and certain fish species from fishing gear, helping to reduce by-catch.

Real-Time Monitoring and Surveillance: The use of electronic monitoring systems, including cameras and sensors, allows for real-time monitoring of fishing operations, providing valuable data to assess and mitigate by-catch.

Biodegradable Fishing Gear: The development of biodegradable or naturally degradable fishing gear materials can reduce the impact of lost or abandoned gear on marine ecosystems, minimizing ghost fishing and by-catch of marine life.

Smart Buoy Technology: Smart buoy systems equipped with sensors and satellite communication can help track fishing gear, monitor fishing activity, and provide real-time data to reduce the risk of lost gear and associated by-catch.

Data Analytics and Machine Learning: Advanced data analytics and machine learning technologies can be used to analyse large datasets related to fishing operations, helping to identify patterns and trends associated with by-catch and informing the development of targeted solutions.

These technological innovations, when combined with effective fisheries management and regulatory measures, can play a crucial role in reducing by-catch and promoting sustainable fishing practices. Ongoing research and collaboration between industry, scientists, and conservation organizations are essential for further developing and implementing these technologies to minimize the impact of by-catch on marine ecosystems.

Overall, investment in R&D for innovative fish harvest technology has the potential to improve the sustainability, efficiency, and environmental impact of fishing activities, while also contributing to the long-term viability of the fishing industry.

Conclusion

Fish harvest technology has come a long way, from traditional methods to modern techniques. While modern fish harvest technology has its advantages, it also has significant impacts on the fishing industry, fish populations, and marine ecosystems. In the past, technological development of fishing gear and methods was aimed at increasing fish production by any means, by increasing efficiency of the fishing gear systems, but in the recent past, due to overfishing, reduction in the total catch of few species groups, overexploitation in the near shore waters, environmental and ecological impacts of fishing, fishing gear development is more oriented towards responsible fishing gear systems which aims at improved size-selective and species-selective properties, minimum impact on the environment and non-target resources, and sustainability of fish stocks. To ensure the sustainability of fish harvest technology, it is important to promote sustainable harvest practices and implement policies and regulations to protect the environment. By doing so, we can ensure that fish harvest technology remains sustainable and contributes to the livelihoods of fishers and the health of our oceans. The way forward is to continue to invest in research and development of new fish harvest technologies that can help reduce by-catch, post-harvest losses and improve the quality of fish and fish products. This will not only benefit the fishing industry but also help reduce food loss and waste, which is a critical issue in today's world.

Further Reading

Boopendranath, M.R., Pravin, P., Thomas, S.N.,Edwin, L., 2009. Handbook of fishing technology. B. Meenakumari (Ed.). Central Institute of Fisheries Technology.

Glass, C.W., Walsh, S.J. and van Marlen, B., 2007. Fishing technology in the 21st century: integrating fishing and ecosystem conservation. ICES Journal of Marine Science, 64(8), pp.1499-1502.

Pravin, P. and Ramesan, M.P., 2013. Advances in harvest strategies: Experiences of CIFT in responsible fisheries. ICT-oriented Strategic Extension for Responsible Fisheries Management, p.57.

Pravin, P., Gibinkumar, T.R., Sabu, S. and Boopendranath, M.R., 2011. Hard Bycatch Reduction Devices for Bottom Trawls: A Review.

10

Demand-Driven Changes in Trawl Catch and Bycatch in India: Suggestions for Trawl Bycatch Mitigation in the Changed Scenario

Dineshbabu A. P.

ICAR-Central Marine Fisheries Research Institute, Kochi – 682 018, Kerala
Email: dineshbabuap@yahoo.co.in

Introduction

Trawl fishing stands out as the predominant method contributing significantly to the marine fish catch, with the trawl fisheries sector representing approximately 40% of the marine fish production on average from 2017 to 2022. This sector plays a pivotal role, contributing over 75% to the marine fisheries export from the country. Critically evaluating trawl fisheries from a sustainability perspective reveals global concerns regarding its impact, particularly as it is identified as a major contributor to bycatch, posing significant ecological challenges. The evolution of export-oriented processing units has led to a shift in species preference among trawl operators in India, favoring cephalopods, ribbonfishes, and threadfin breams due to their substantial export market demand. To align with changing market demands, adjustments in trawl gear design are imperative to efficiently target the preferred species. The primary bottom shrimp trawl gear has diversified into five major types in the region: shrimp trawl, perch trawl (bottom trawls), cephalopod trawl (column trawl), and ribbonfish trawl (pelagic trawl).

Genesis and development of trawl fishery

The inception of trawl fishery along the Indian coast dates back to 1953 when a limited number of boats were introduced by the Indo-Norwegian Project. Over the ensuing six decades, notable changes have transpired, encompassing alterations in the operational depth, duration of fishing trips, and species composition of landings. In the early 1960s, the operational depth of trawl fishing activities ranged between 10-20 meters, expanding to 30 meters in 1967-70. Subsequently, during the period of 1970-80, the operational depth increased to 40 meters, reaching 55 meters in 1980-85. Notably, during this timeframe, night trawling and multi-day trawling practices were introduced. Between 1985 and 1995, the operational depth escalated to 100 meters, and by the year 2000, it further extended to 150 meters. The trawl fishery has encountered challenges in terms of low profitability attributed to diminished catch per unit and heightened fuel expenses. This scenario has given rise to intense competition within the trawl fishery sector, manifesting in the adoption of augmented engine power, varied trawl designs, enhanced storage capacity, and prolonged trip durations.

Ever since marine fishery operations were extended beyond territorial waters of respective states, the need was felt for incorporating spatial component of fishing effort and catch for formulating marine fishery policies. Spatio-temporal analysis of changing scenario of fishing operations in GIS platform with catch and effort data collected from fishermen, highlight the need for the change in approach in fisheries management.

Spatial analysis of fishing operation

Spatial study (Fig.1) of trawl fishing operations from different states were carried out across the State territorial water borders. Marine fishing scenario have undergone significant changes in terms of operational depth and area. These changes should be reflected in future fishery management plans for better regulation of the fishery and for sustaining many of the fishery resources. It was understood that trawl fishing is being carried out in much organised manner with most of the trawler operating groups keeping very elaborative log sheet, which is kept in high confidentiality, which is shared only with their own group members. Prevalence of this practices, can be considered as promising opportunity for bringing in spatiotemporal suggestions to sustain the production.

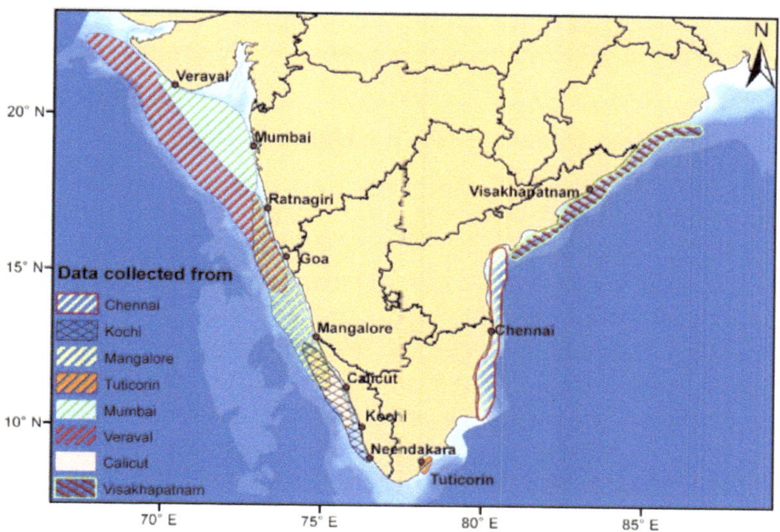

Fig. 10.1: Trawl fishing grounds mapped (2017) from trawl fishery centres along Indian coast

Engine Evolution and Trawling Advancements

Significant transformations in trawling operations occurred during the specified periods, marked by notable changes in engine specifications. Bull trawling was introduced in 1972, and by the 1980s and 1990s, engine horsepower (HP) increased to the range of 122-145, accompanied by an increase in boat size to 50-60 feet. Fishing days also experienced an upswing, ranging from 6-10 days, with operational depths extending to 30-45 meters.The 2000s witnessed major advancements, characterized by an operational depth increase to 100-120 meters, a fishing duration of 10 days, and a further rise in engine HP to 185. During 2006-07, engine HP escalated to 245, with operational depths reaching 150 meters. In 2008-09, trawlers measuring 58-60 feet were equipped with 350 HP Weichai and Uchai engines, facilitating an operational depth increase to 200 meters, and fishing durations ranging from 10-14 days. The subsequent shift to steel trawlers in the 60-70 feet range with high HP engines continued from 2010 onwards, allowing for operational depths up to 200 meters and fishing durations spanning 10-14 days. Examining the average fishing hours per trawler reveals a substantial increase over the last two decades, rising from 7 hours in 1990 to exceeding 33 hours by 2012. The evolution of trawl nets and operational techniques has undergone rapid transformation since 2000, coinciding with a substantial shift from traditional wooden trawlers to steel trawlers on a large scale. It is noteworthy that these modifications have occurred at an accelerated pace, lacking a scientific rationale, management regulations, and consideration for ecosystem dynamics. The absence of a structured framework has contributed to the unsustainable evolution of trawl fishing practices.

Transition from Bottom to Semi-Pelagic/Pelagic Trawling

The advent of high engine power enabled a shift from traditional bottom trawling to semi-pelagic/pelagic trawling. Various trawl designs, including two seam trawls, four seam trawls, six seam trawls, bulged belly trawls, high opening trawls, and large mesh trawls, have been conceived and developed in India to enhance fishing efficiency and adapt to evolving requirements within the industry. This transition became evident through the analysis of trawl landing data over decades, specifically in 1992, 2002, 2012, and 2022. Trawl fishery catch nearly doubled between 1992 (10,92,472 tons) and 2022 (20,17,492 tons). The analysis categorized fish landed based on ecosystem preference, distinguishing Demersal species (bottom-dependent fin fishes and shellfishes), Column species (exhibiting free movement within the water column), and Pelagic fishes (preferring the upper strata of the ecosystem).

Species Composition Shift in Trawl Landing

From the year 2000 onward, a notable transition in the species composition of trawl landing was observed, particularly characterized by a significant increase in column fishes such as cephalopods and threadfin breams. The surge in demand for cephalopods from European countries, high-value single market demand for ribbon fishes from China, and the preparation demand for "surumi" from threadfin breams, particularly *Nemipterus randalli*, prompted fishermen to modify nets and fishing techniques to enhance the economic feasibility of trawling operations. During this period, fishing voyages extended to more than 10 days of operation, indicating a shift towards pelagic and semi-pelagic trawling. Data analysis of four years, representing four difference decades, 1992, 2002, 2012 and 2022, also convincingly supports the observation of increasing pelagic and semi-pelagic trawling operations.

Change Analysis of Trawl Fisheries southwest coast of India

The investigation in Mangalore revealed that, on average, 50% of the available Marine Demersal Fishery (MDF) area was subjected to trawling activities from 2008 to 2018. The minimum trawled area occurred in 2011, while the maximum was observed in 2017. The increase in fishing days and engine capacity over this period contributed to the expansion of trawling areas. This analysis sheds light on the evolving dynamics of trawl fisheries in Mangalore, emphasizing the correlation between fishing practices, engine capabilities, and the extent of trawling activities. The observed trend suggests a noteworthy increase in the utilization of the available marine demersal fishery area, reflecting adjustments made in response to demand-driven modifications in gear design and operational strategies.

When comparing the trawled areas it is observed that in 2008, when bottom trawling was done there was a gap in 100 m depth, which was due to the presence of rocky patch (seamount) in that area which could not be trawled. However in 2018, when the pelagic trawl operation is widely done and the un-trawled area were also trawled.

Introduction of pelagic trawling happened during 2015 which has resulted in the change of species composition in multiday trawlers. The percentage of pelagic fishes in commercial as well as low value bycatch (LVB) has increased during the recent times when compared to 2008. The increase in *Decapterus russelli* and *Rastrelliger kanagurta* and other pelagic species were observed in catches from 2017 onwards when compared to 2008. The demersal contribution to the LVB has also decreased during the later period (Fig. 2 & 3).

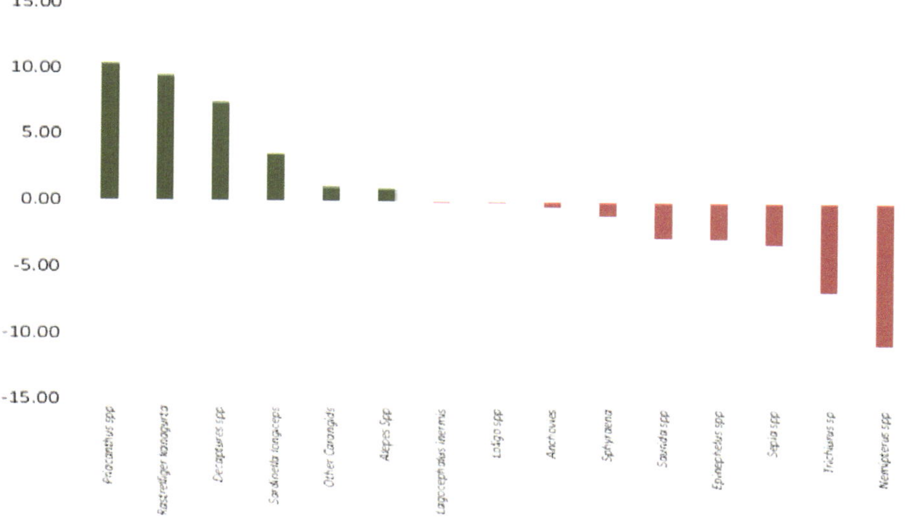

Fig. 10.2: Comparison of percentage contribution of commercial species between 2008 and 2018

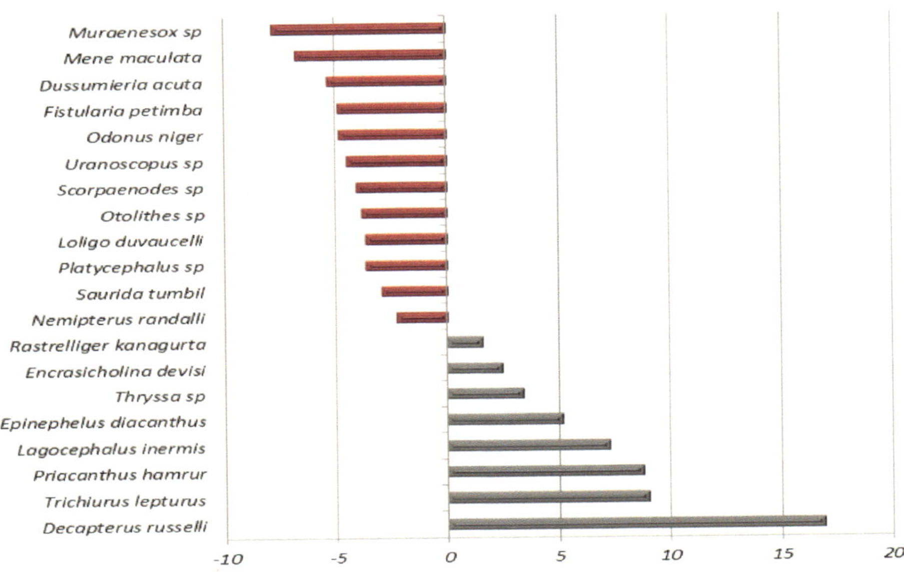

Fig. 10.3: Comparison of species contribution in LVB between 2008 and 2018

Change Analysis of Trawl Fisheries from Northwest Coast of India

Changes in Species Preference

Trawl operations were initially introduced to exploit the abundant shrimp and demersal fish resources. However, with the establishment of export-oriented processing units, the demand for other species with good export market potential, such as ribbonfishes, increased significantly among the fishers in Gujarat. In the past decade, several processing units obtained approval from the European Union (EU), leading to cephalopods (squid and cuttlefishes) emerging as preferred species for boat operators in Gujarat. Additionally, *Aluterus monoceros,* a species previously discarded, gained value as a high-value export commodity in the region.

Changes in Design of Trawl Gear

To meet the evolving demand patterns, the design of trawl gear has undergone alterations to efficiently harvest preferred species. The primary bottom shrimp trawl gear has diversified into five major types: shrimp trawl, perch trawl (bottom trawls), cephalopod trawl (column trawl), ribbonfish trawl (pelagic trawl), and Acetes trawl (single day operation). These gears differ primarily in dimensions, particularly in head rope length and mesh configuration in wing and belly areas. Ribbonfish trawls are the largest in dimensions, followed by cephalopod trawls. Despite these differences, the cod-end mesh size remains similar in all trawl gear types, except for Acetes species trawl, which features a significantly smaller cod-end mesh size. The towing speed of each trawl gear is customized according to the target species, reflecting the adaptability of gear design to specific harvesting requirements.

Vertical Diversification of Trawl Effort

With the demand shifting significantly towards column and pelagic resources, the trawl effort has been strategically diverted towards pelagic or column trawling operations. Bottom trawling now accounts for only one-third of total fishing hours, while the remaining two-thirds are dedicated to ribbonfish (pelagic) or cephalopod trawling (column). A retrospective study based on questionnaires revealed that in 2005, the proportion of bottom trawling was approximately 70%. Since then, pelagic and column trawling have exhibited an average growth rate of around 3% per year, indicating a deliberate shift in trawl effort towards more economically viable and ecologically sustainable fishing practices aligned with market demands.

Low-Value Bycatch and Discards Landings and Utilization

The catch rates of low-value bycatch (LVB) and discards exhibit seasonal variations, with higher discard rates during peak fishing seasons, when the availability of high-value fishes is abundant (post-monsoon season). Conversely, during lean seasons, discard rates are lower, and a major proportion is landed as low-value bycatch. In the last decade, however, the proportion of discards has reduced. Several previously discarded species, including puffer fish, clupeids, leiognathids, and juveniles of other fishes, are now utilized as raw materials in fish meal factories (of poorer quality) or even for surimi production (for juveniles of edible fishes). Consequently, during lean seasons, the landing of these resources allows fishers to achieve breakeven or marginal profits. Additionally, with a significant portion of trawl effort directed towards column or pelagic realms and a major share of bottom trawl effort focused on off-bottom perch trawling, the landings and discards of non-target invertebrates like gastropods, box crabs, and echinoderms have substantially reduced along the Gujarat coast in the last two decades.

Initiatives in Bycatch Mitigation in India

General bycatch conservation and management principles have been developed based on a multidisciplinary approach, involving a larger scale of people, and incentivization for bycatch mitigation. In India, various measures have been implemented to mitigate bycatch and ensure the sustainability of marine fisheries. These include seasonal closure of fisheries, implementation of Minimum Legal Size (MLS), alterations to gear design such as mesh size optimization, installation of bycatch reduction devices (BRD), juveniles and trash fish excluder devices. These changes have been experimented with on a regional and national scale, with varying rates of success in reducing low-value bycatch.

Seasonal closure of fisheries has proven highly successful in reducing fishing pressure, especially since the closure aligns with the spawning season of fishes. Experiments from Mangalore have expanded the scope of regional closures by mapping juvenile grounds of commercially important species and suggesting these areas for regional and seasonal closures. Identifying areas of high bycatch occurrences and juvenile assemblages through regional closures and implementing MLS are essential steps. Avoidance of such areas during prominent seasons can significantly reduce bycatch in trawl fishery. Mapping juvenile and spawner abundance along the Indian coast using GIS provides a database for spatial and temporal closures or restrictions in trawl fishery, enabling more informed and targeted conservation measures. GIS-based resource mapping serves as a valuable tool for policymakers, allowing transparent decision-making by weighing each fishing ground in terms of commercial value and juvenile abundance.

Resource maps, utilizing GIS, contribute to awareness programs by providing stakeholders with visual information on seasonal and fishing ground-wise distributions of juveniles and commercial fishes. These maps aid in identifying critical fishing grounds where seasonal and spatial closure of trawl fishery can be implemented to improve long-term fishery production. The examples of mapping grounds of peak spawning and juveniles' abundance of commercial species serve as an informativon base for suggesting seasonal and spatial closure of the fishery, contributing to sustainable and economically viable fisheries management (Fig. 4 & 5).

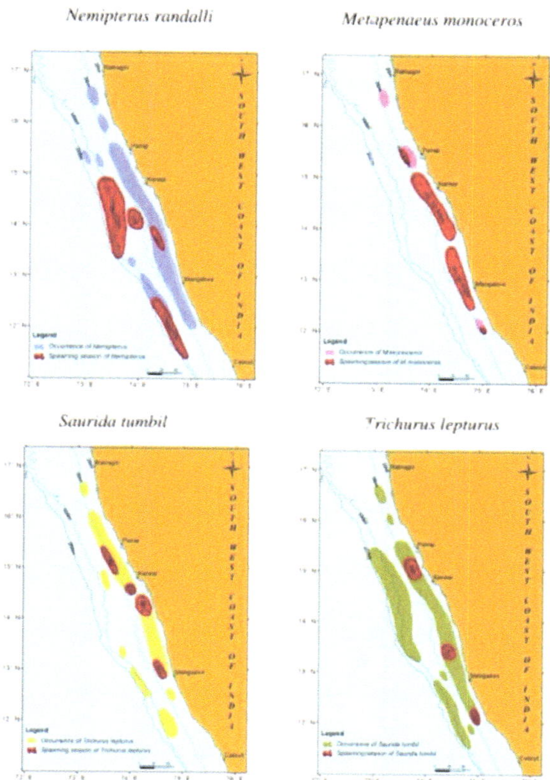

Fig. 10.4: Seasonal peak spawning grounds identified for *N. randalli, M. monoceros*, *S. tumbil* and *T. lepturus* based on the GIS studies. Peak spawning months and area of occurrence is indicated in the map

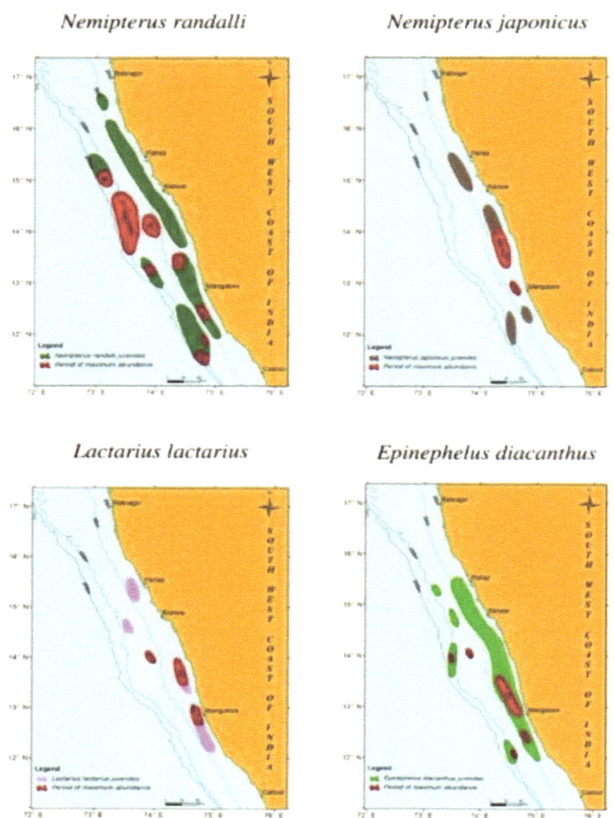

Fig. 10.5: Seasonal juvenile abundance grounds identified for *N. randalli, N. japonicus, L. lactarius* and *E. diacanthus* based on the GIS studies. Months of juvenile abundance and area of occurrence is indicated in the map.

Impact of MLS implementation on LVB

During the period of study, Minimum legal size (MLS) at the landing centers was strictly implemented in Kerala from 2017, and this provided an opportunity for analyzing the impact of MLS implementation on LVB landings and its juvenile composition. Data from the Malabar region (Beypore and Puthiyappa landing centers) were analyzed (Fig.6). Before implementation of MLS, LVB landings of multi-day trawl netters (MDTN) from Calicut were about 35% of the landing, of which 43.7% were formed by juveniles of commercially important species (in terms of numbers they formed 53%). After the effective implementation of MLS and strict monitoring by the Marine Enforcement Squad of Government of Kerala starting 2018, the LVB from MDTN reduced to 15% of the landing, and more significantly this reduction was chiefly due to the reduction in juvenile landings in the LVB, which was found to have reduced to 16.3% of the LVB landings (20.8% by number).

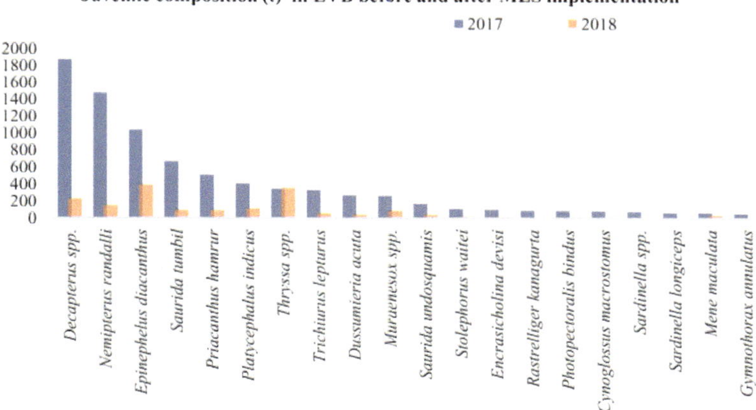

Fig. 10.6: Changes in juvenile composition (t) in trawl fishing centres of Calicut during the years 2017 and 2018

It can be assumed that after the introduction of MLS in marine fisheries, the landing of the LVB and juvenile percentage in LVB reduced significantly. Evaluating the MLS introduction results from different case studies, indicates that MLS implementation was a very positive step toward reducing juveniles in the bycatch, however, it has to be implemented in conjunction with mesh size regulation, or else it may lead to more discards from the trawlers.

Promising Case Studies of Low-Value Bycatch Conversion into Edible Protein

Many species initially considered low-value bycatch became highly commercially valuable as market opportunities were explored. Examples include pufferfishes and triggerfishes. Market demand led to the conversion of pufferfishes, once considered a menace and low-value bycatch, into a highly valued commercial species, particularly in the domestic and export markets.

Way Forward

The trawl fishery plays a crucial role in the livelihoods of millions of primary, secondary, and tertiary stakeholders. To ensure the well-being of the country from a holistic perspective, it is imperative to adopt a balanced approach that takes into account both economic and ecological factors. Numerous efforts have been made to analyze the bycatch composition across India, raising concerns about the disproportionately high percentage of juveniles, particularly those of commercially valuable fish species, in the landings of the LVB.

Given the information presented, there is a compelling need for a significant deviation in trawl fishery management. This contradicts the global concern of bottom destruction, as Indian trawlers have alleviated pressure on bottom species due to increased demand for column and pelagic fishes. State-of-the-art technologies in engine capacity and gear modifications have enabled faster trawling, contributing to the reduced impact on bottom flora and fauna within the trawling ecosystem. Although this has lessened pressure on the bottom, high-speed engines are capable of rapidly reaching shoals, resulting in an increased composition of juveniles, which is a matter for concern. Therefore, it is crucial to conduct further research on reducing juvenile bycatch through the effective implementation of Minimum Legal Sizes (MLS). The emphasis should be on developing juvenile avoidance methodologies instead of compelling trawlers to discard juveniles to comply with regulatory measures. To achieve effective juvenile avoidance, it is essential to conduct mapping of juvenile grounds using GIS technology. Additionally, the compulsory implementation of a Vessel Monitoring System (VMS) is necessary to enforce avoidance of fishing grounds identified for juvenile protection.

Further Reading

Dineshbabu, A.P., Radhakrishnan, E.V., Thomas, S., Maheswarudu, G., Manojkumar, P.P., Kizhakudan, S.J., Pillai, S.L., Chakraborty, R.D., Josileen, J., Sarada, P.T. and Sawant, P.B., 2013. Appraisal of trawl fisheries of India with special reference on the changing trends in bycatch utilization. Journal of the Marine Biological Association of India, 55(2), pp.69-78

Dineshbabu, A.P., Thomas, S. and Radhakrishnan, E.V., 2012. Spatio-temporal analysis and impact assessment of trawl bycatch of Karnataka to suggest operation-based fishery management options. Indian Journal of Fisheries, 59(2), pp.27-38

Dineshbabu, A.P., Thomas, S. and Shailaja, S., 2016. Impact of trawling in Indian waters-A review. Fishery Technology, 53, pp.263-272

Dineshbabu, A.P., Thomas, S. and Vivekanandan, E., 2014. Assessment of low value bycatch and its application for management of trawl fisheries. Journal of the Marine Biological Association of India, 56(1), pp.103-108

Dineshbabu, A.P., Thomas, S., Josileen, J., Sarada, P.T., Pillai, S.L., Chakraborty, R.D., Dash, G., Chellappan, A., Ghosh, S., Purushottama, G.B. and Kumar, R., 2022. Bycatch in Indian trawl fisheries and some suggestions for trawl bycatch mitigation. Current Science, 123(11), pp.1372-1380

Dineshbabu, A.P., Thomas, S., Radhakrishnan, E.V., Sreedhara, B., Muniyappa, Y., Kemparaju, S. and Nataraja, G.D., 2011. Mapping of fishery resources in trawling grounds along the Malabar-Konkan coast. Marine Fisheries Information Service, (210), pp.1-12

Mahesh, V., Benakappa, S., Dineshbabu, A.P., Naik, A.K., Vijaykumar, M.E. and Khavi, M., 2017. Occurrence of low value bycatch in trawl fisheries off Karnataka, India. Fishery Technology, 54(4), pp.227-236

Mahesh, V., Dineshbabu, A.P., Naik, A.S., Anjanayappa, H.N., Vijaykumar, M.E. and Khavi, M., 2019. Characterization of low value bycatch in trawl fisheries off Karnataka coast, India and its impact on juveniles of commercially important fish species. Indian Journal of Geo Marine Sciences, 48 (11):1733–1742

11

Sustainable Development of Fish Harvest and Post-Harvest Sector Government Initiatives in India

L. Narasimha Murthy

Chief Executive, National Fisheries Development Board, Department of Fisheries Government of India, Hyderabad- 500 052, Telangana

Email: drlnmurthy@gmail.com

Introducvtion

The Indian fisheries sector is programmed for sectoral transformation through Pradhan Mantri Matsya Sampada Yojana (PMMSY) introduced by the Government of India. The main motto of PMMSY is 'Reform, Perform and Transform' in the fisheries sector. Launched in 2020, the flagship scheme is dedicated for focused development of Indian fisheries sector with an estimated investment of Rs 20050 crore for a period of 5 years from FY 2020-21 to FY 2024-25 (Figure 1). It is an umbrella scheme with two separate components namely (a) Central Sector Scheme (CS) and (b) Centrally Sponsored Scheme (CSS) (Figure 2).

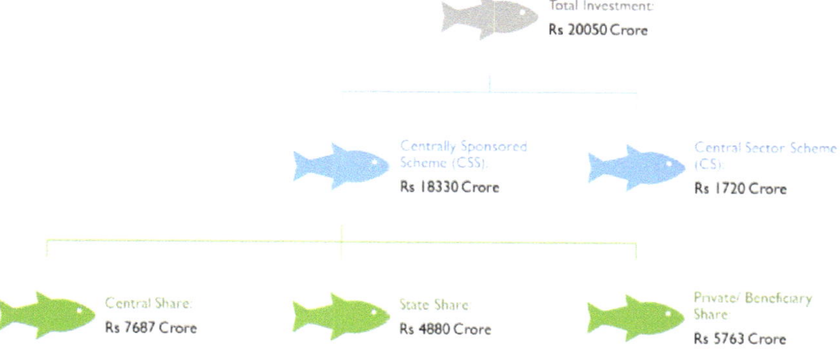

Fig. 11.1: PMMSY Investment Break-Up

Implementation of PMMSY addresses critical gaps in fish production and productivity, postharvest infrastructure, domestic fish consumption, traceability and fishers' welfare. It is envisaged to create 55 lakhs employment opportunities in the sector (15 lakhs in direct fisheries activities and 40 lakhs in allied activities) and to provide livelihood and nutritional support during fish ban/lean period to 6.77 lakh families. During the period of FY 2020-21 to 2023-24 (till October 2023), 47.19 lakhs employment opportunities has been generated through PMMSY. A new Central Sector Sub-scheme under PMMSY with targeted investment of Rs 6000 crore with the objective to enhance further the earnings and incomes of fishermen, fish vendors and micro & small enterprises engaged in fisheries sector, was announced during the Union Budget presentation for FY 2023-24.

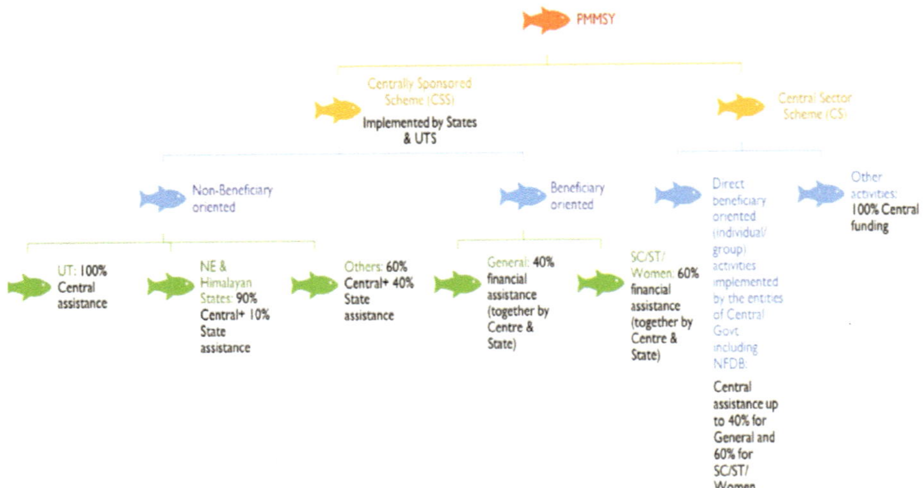

Fig. 11.2: PMMSY Funding Pattern

Developments in Fish Harvest Sector

PMMSY promotes 'Make in India' in fish harvest sector. The scheme supports modernization of fishing vessels, promotes the use of low-cost indigenous fishing vessels, mother vessels etc. Realising the importance of fishing harbours & fish landing centres as a point of convergence between production and trade, its modernization and development are prime area of focus by the Government. Till October 2023, 463 deep sea fishing vessel, 1143 proposals for upgradation of existing fishing vessels, 255 proposals for construction of Bio-toilets in mechanized fishing vessels and 22 integrated modern fish landing centers are approved under PMMSY. There are currently 107 'Fishing Harbours and Fish Landing Centres' being developed and modernized under

Blue Revolution, Fisheries and Aquaculture Infrastructure Development Fund (FIDF) and PMMSY.

PMMSY aims at doubling the fisheries export Rs. 1 lakh crore by 2025. In FY 2022-23, the Indian fisheries sector produced 175.45 lakhs metric tonnes of fish and contributed Rs. 63969.14 crore to the Indian economy through the export of 1.74 million metric tonnes of seafood. One of the key implementation strategies of PMMSY is 'cluster/area-based approach'. This is mainly being implemented in species diversification (Tilapia, Pangasius, Sea Bass, Scampi and Mud crab), value addition of fish products, infrastructure and logistics support, marketing and promotion of brand India etc. The Government is keen to develop a functional and sustainable cold chain throughout the country to prevent/minimise the post-harvest losses in fisheries sector and to ensure nutritional security. Towards realising this, it has supported funding of 563 ice plant/ cold storages and 25193 fish transportation facilities in total till October 2023, including 9282 bicycles with ice box, 1138 insulated trucks, 9858 motor cycles, 3640 auto rickshaws, 294 refrigerated trucks to the beneficiaries through PMMSY.

Though domestic fish marketing in India holds a huge potential, it is still highly unorganized and unregulated. The improvement in fish marketing system and distribution would not only reduce the demand-supply gap of fishes across country, but would also contribute to food and nutritional security of a vast majority of resurgent middle-income population, since fish contributes to ensuring domestic food and nutritional security. India registered a per capita yearly consumption of over 13 kg in 2022-23. In addition, more varieties of fish are available for consumption by households through expanding market networks. The current per capita consumption, though lower as compared to the international estimate (OECD & FAO) of 20.5 kg per capita (2019-20) has a high potential to catch up to the predicted level of 21.4 kg per capita by 2031. Monthly consumption of fish per household has shown a quantum leap in ten years from 2.66 kg in 2011-12 (NSS 68th Round) to 4.99 kg in 2022-23 (NCAER). Recognising the potential of domestic marketing, the Government funded 9 State of art wholesale markets, 188 fish retail market, 108 value added enterprise units and 6542 fish kiosks including ornamental fish kiosks through PMMSY till October 2023.

Entrepreneur Model of National Fisheries Development Board

NFDB undertakes need based beneficiary-oriented fisheries development activities as 'Entrepreneur Models' under the CS Component of PMMSY. A minimum of Rs. 100 crores is earmarked for the scheme under PMMSY, for a period of five years from 2020-21 to 2024-25. NFDB gives financial support of up to 25% of the total project cost for the beneficiaries belonging to General category and up to 30% for those belonging to SC/ST/Women category with a ceiling of Rs. 1.25 crores and Rs. 1.50 crores per project respectively. So far, 38 projects have been sanctioned by NFDB under Entrepreneur Model with the total outlay of Rs. 114.034 crores and eligible subsidy of Rs. 30.65 crores, of which 8 are from post-harvest sector. These 8 projects include setting up of pre-processing units, high-end processing units such as Individual Quick-Freezing unit, cold storages, fish value addition unit including fish drying unit, fish markets, fish trade centre and cold chain development through refrigerated and insulated vehicles. Some of the activities under post-harvest sector eligible to be supported under 'Entrepreneur Model' are hub and spoke models in urban areas for promotion of domestic fish consumption, leveraging the market yards in rural and semi/peri urban areas for promotion of fish domestic consumption, post-harvest infrastructure including cold chain and marketing, etc.

Opportunities for Scientific Community under PMMSY

PMMSY is a great platform for the scientific community (a) to establish/carry out their research outcomes/ inventions in a larger scale and (b) to expand their basic research to applied research for bridging the existing gaps in the fisheries sector. The sub-components/ activities with 100 % central funding under the Central Sector Scheme of PMMSY (Table 1) have the scope to be utilised by the scientific community.

Table 1: Sub-components/ activities with 100% central funding under the Central Sector Scheme of PMMSY

1	Genetic improvement programmes and Nucleus Breeding Centers (NBCs)
2	Innovations and Innovative projects/ activities, Technology demonstration including startups, incubators and pilot projects
3	Training, Awareness, Exposure and capacity Building
4	Aquatic Quarantine Facilities
5	Modernization of fishing harbours of central government and its entities
6	Support to NFDB, Fisheries Institutions and Regulatory Authorities of Department of Fisheries, Government of India and need based assistance to State Fisheries Development Boards
7	Support for survey and training vessels for Fisheries Institutes including dredger TSD Sindhuraj owned by the DoF and GoI
8	Disease Monitoring and Surveillance Network
9	Fish data collection, fishers' survey and strengthening of fisheries database
10	Support to security agencies to ensure safety and security of marine fishermen at sea

11	Fish Farmers Producer Organizations/ Companies (FFPOs/Cs)
12	Certification, accreditation, traceability and labelling
13	Administrative Expenses for implementation of PMMSY (to meet expenses of both for CS and CSS components)

Source: PMMSY Operational guidelines in English. Department of Fisheries, Government of India. pp. 55-73. https://pmmsy.dof.gov.in/new-download

Innovations and innovative activities related to fisheries and aquaculture including pilot projects, demonstrations, etc. are encouraged and supported by NFDB with a ceiling of Rs 3 crores (evaluated on case-to-case basis) under the sub-component 2 of CS component of PMMSY. Startups in fisheries, fisheries incubators centres, new advances in capture fisheries, innovative approaches to address malnutrition are a few which are supported under this sub-component. Likewise, special focus is accorded under PMMSY for training, awareness, exposure and capacity building of fishers, fish farmers, fish workers/vendors and officials.

Fisheries and Aquaculture Infrastructure Development Fund (FIDF)

To encourage State and UT Governments, private entrepreneurs, fish farmers, institutions etc. in creation of fisheries infrastructure facilities, Government of India created Fisheries and Aquaculture Infrastructure Development Fund (FIDF) with an estimated fund outlay of Rs 7522.48 crore. NFDB, Hyderabad is the Nodal Implementing Agency for FIDF Scheme. Establishment processing units, labs, harbours, training centres, cold storages, fish markets, etc., introduction of deep-sea fishing vessels, fish transport facilities etc. are supported under this fund.

New Initiatives by the Government in Harvest and Post-Harvest Sector

1. *Vessel Communication and Support System on Marine Fishing Vessels:* Satellite-based Vessel Communication and Support System Developed by ISRO, Department of Space is being installed on one lakh fishing vessels for ensuring safety and security of fishermen at sea, enabling them to stay connected with their families and seeking assistance during cyclones and storms or incase fishing near international borders.
2. *Use of green fuels (LPG) for outboard engine boats to reduce GHG emissions:* In line with India's commitment towards reducing its carbon footprints, a pilot project on use of cleaner alternative fuels in motorised fishing vessels has been initiated. Pilot trials conducted by M/s Surya Marines in association with CIFNET and MATSYAFED, Kerala concluded that LPG kits are suitable to replace the petrol & kerosene-based motor engines as they have low running costs, reduced

environmental impact, longer life due to less wear & tear of the internal parts, etc.

3. *Phase Change Material:* In the current scenario, the cold chain management in the fisheries sector is dominated by the usage of thermocol boxes with ice for chilling and storage of the fish by the fishers/farmers/ fish vendors/retailers. However, availability of quality crushed ice and the costs involved in purchasing ice, insulated box usage are the main constraints faced in the fisheries sector which are contributing to the more post-harvest losses. Phase Change Material (PCM) is a material innovation that serve as an alternative for the ice usage which may cut down the recurring costs on the fish storage. Pre-cooled PCM packed in specially designed thermal panels/cassettes, that releases/absorbs sufficient energy at phase transition thus resulting in cooling of the other material when used in insulated boxes eliminates/reduces the amount of ice used for fish chilling during fish vending. This enhances shelf life of fish, fish portability and product quality in addition reduces energy consumption. NFDB has conducted a techno-economic feasibility study of the use of PCM boxes.

Way forward for Harvest and Post-Harvest Sector

1. Introduce better deep-sea fishing vessels and conversion of existing vessels and gears to use green and safe energy as per national policy
2. Fill the large infrastructure gaps in fisheries sector by developing fishing harbours/ fish landing centres, creating cold chain infrastructure facilities such as ice plants, cold storage, fish transport facilities, fish processing units, fish markets, etc.
3. Promote research to reduce harvest and post-harvest losses in fisheries sector and to develop methods to utilize fishery waste into functional and value-added products and by-products
4. Improve storage stability of the fish and fishery products especially during the last mile delivery

Further Reading

https://pmmsy.dof.gov.in/
https://nfdb.gov.in/
https://dof.gov.in/

12

Emerging Threats to Aquatic Ecosystem Health

Pramod Kumar Pandey* and Amit Pande

ICAR-Central Institute of Coldwater Fisheries Research, Bhimtal-263139, Uttarakhand
**Email: pkpandey_in@yahoo.co.uk*

Introduction

An aquatic ecosystem is a system of interactions between living beings and the environment they inhabit. It includes the aquatic plants and animals that dwell in water and the physical and chemical factors that impact the aquatic ecosystem. Water, a physical environmental component, makes up the aquatic ecosystem. The condition of the water body determines whether plants and animals can grow and whether or not they will survive. The physical characteristics of these habitats, like pH, temperature of the water, and the amount of light penetrating the water, are all significant physical factors that affect the plants and animals of an ecosystem. The concentrations of nutrients, temperature, the flow of water, and availability of shelter all have direct and immediate effects on the living organisms that inhabit a particular environment. Species abundance and diversity are affected by interactions between aquatic organisms, such as competition for resources (like food and habitat) and predation. Therefore, understanding the fundamental components of aquatic ecosystems and the interactions between living organisms and their environment is required to measure human impacts. The 17 Sustainable Development Goals (SDGs) represent a pressing demand for action from all developed and developing nations in a worldwide collaboration. They acknowledge the necessity of simultaneously addressing poverty and other forms of deprivation alongside implementing strategies to enhance health and education, diminish inequality, and stimulate economic growth. In this direction, clean water and sanitation (SDG 6), climate action (SDG13) and life below water (SDG 14) need to be well addressed to achieve zero hunger (SDG 2). Fish is the most important source for millions around the world the quality and quantity of water need to be conserved, which is an important constituent of aquatic ecosystems.

An aquatic ecosystem can be a part of a freshwater or marine environment. Freshwater aquatic ecosystems include lentic, lotic, swamps, and wetlands, where as the marine ecosystem comprises the ocean, estuaries, coral reefs, and coastal ecosystems. All living beings in an ecosystem are affected by abiotic and biotic factors. Abiotic factors are non-living environmental components that influence the biotic components, including organisms or an ecosystem's living components. The presence of living beings and the biological by-products they produce influence the composition of an ecosystem. The health of aquatic ecosystems is important for the survival of aquatic life forms. However, contaminants that human interventions have frequently introduced have led to serious consequences. These contaminants have disturbed the biodiversity and destroyed the aquatic habitats.

Aquatic ecosystem health

Maintaining ecological structure, processes, functions, and resilience within the range of natural variability is the hallmark of a healthy aquatic ecosystem. Well-functioning ecosystems offer numerous advantages. A healthy ecosystem consists of indigenous species, with natural processes fully operational and functioning. Occasionally, some occurrences harm an ecosystem and the individuals who depend on it. These effects can be alleviated by implementing well-informed and scientifically grounded management strategies. Healthy ecosystems benefit human populations by providing support, resources, regulation, and cultural connections. A robust aquatic ecosystem can enable watershed security, which supports all forms of life and acts as a protective barrier for human populations from the most adverse types of weather. However, the world's water resources are increasingly threatened by climate change, population growth, and pollution. As the global population grows, a persistent challenge is accessing enough water to meet humanity's needs while preserving the integrity of aquatic ecosystems.

Monitoring aquatic health

Water health monitoring measures water quality, quantity, and biological indicators. It is crucial for enhancing comprehension of the aquatic ecosystem and informing sustainable water management decisions. Monitoring activities encompass gathering, examining, understanding, and disseminating information. The data collected by monitoring programs or the baseline data is utilized to assess the historical and current condition of water and aquatic ecosystems. Collecting more baseline data each year enhances our capacity to comprehend current environmental conditions and predict future ones. Baseline data also enable the evaluation of cumulative effects, establishment of thresholds, and formulation of mitigation strategies for different water users. Cumulative impacts refer to the alterations in the biophysical, social, economic, and cultural surroundings resulting from previous, ongoing, and anticipated actions. Knowing the present water condition and anticipated water consumption aids in identifying and assessing patterns. Additionally, it enables the ability to forecast the consequences of upcoming development endeavours on water and aquatic ecosystems.

Aquatic ecosystem health indicators evaluate the water and aquatic ecosystem's condition and ascertain whether natural or anthropogenic environmental disruptions undermine ecosystem processes. Indicators of aquatic ecosystem health encompass various parameters such as water and sediment quality, water quantity, flow measurements examining the physical and chemical characteristics of water and sediments, and biological indicators assessing the population, health, and habitat of aquatic plant and animal species.

Threats to an aquatic ecosystem

Water, known as the vital force of life, sustains all living beings and makes up 70 % or more of an organism's mass. The Earth stands out from other celestial bodies due to its possession of water, which sustains life. Aquatic environments are renowned for harbouring a wide array of organisms. Approximately 4.3 million cubic kilometres of freshwater must be divided among the population of Earth. Water is highly versatile compared to other liquids on Earth and is commonly referred to as a "universal solvent" due to its ability to dissolve various chemical compounds or suspend particulate organic matter. The suitability of water for use, whether it is sufficiently clean or excessively polluted, depends on the actions taken by the users. Aquatic ecosystems face several threats, like climate change, habitat damage, pollution, disease, and alteration of sediment loads. These issues have serious implications for the aquatic life.

Climate change

Climate change is arguably the greatest threat to ocean health. It makes oceans hotter, promotes acidification, and reduces dissolved oxygen levels. Climate change has diverse impacts on both the quality and quantity of water. These impacts include alterations in precipitation patterns, the timing and duration of ice formation and melting, as well as ground subsidence and slumping caused by the melting of permafrost. Each of these alterations has the potential to impact aquatic ecosystems. By monitoring indicators of aquatic ecosystem health, it becomes possible to ascertain the impacts of these changes. Climate change a significant issue, can potentially modify the temperature, precipitation, and wind patterns and cause rising sea levels. Surface water temperature elevation can cause stratification in water columns, impeding the oxygenation of the lower layers of a water body, leading to the development of hypoxic and even anoxic conditions in eutrophic systems.

Moreover, the increase in temperatures and the additional nutrients can create favourable conditions for the proliferation of detrimental algal blooms. Elevated water temperature promotes the proliferation of dinoflagellates, which are responsible for producing numerous natural toxins. Seafood tainted with algal toxins leads to syndromes of seafood poisoning, primarily caused by neurotoxins that disrupt the nervous system by interfering with the transmission of nerve impulses. The increased surface water temperatures in tropical oceans caused by El Niño significantly impact the growth of algal blooms. Changes in precipitation patterns have a significant impact on the occurrence of eutrophication. Climate change can cause shifts in wind patterns, impacting the circulation and mixing of oxygen in oceans and coastal areas.

Eutrophication

Aquatic pollution poses a significant problem and grave threat to aquatic life. The eutrophication of water bodies has resulted in the degradation of several species' habitats due to algal blooms. In an aquatic ecosystem, eutrophication occurs due to a gradual rise in the concentration of nutrients, such as phosphorus and nitrogen. It degrades the water quality in the water bodies, causing disruption to aquatic ecosystems and having severe consequences for aquatic life. Eutrophication has intensified over centuries, and human activities have accelerated, creating multiple areas devoid of biological activity. The accumulation of nutrient-rich soil runoffs, sludge, domestic and industrial effluents, and sewage in water bodies due to eutrophication enriches the aquatic environment, leading to hypoxic conditions that disrupt aquatic life. Algal blooms in water bodies can modify the amount of light that enters the water and the temperature and oxygen levels. For many years, eutrophication was regarded as an irreversible phenomenon. Nevertheless, implementing strategies to control human nutrient emissions and mitigate nutrient load in aquatic resources has partially reversed the situation.

Plastic pollution

Plastic is one of the most serious problems affecting our water bodies. Plastic accumulates in the environment, and synthetic plastic products have seriously affected wildlife and their habitats. It is alarming that several million tons of plastic products end up in the oceans every year, and discarded plastic litter is the major constituent. Oceanographic studies have revealed that more than 5.25 trillion individual plastic particles weighing roughly 244,000 metric tons float on or near the surface. Research has revealed that more than 14 million metric tons of microplastic particles are resting on the ocean floor,creating microplastic hot spots. Plastic pollution can kill marine fish and mammals through ingestion and entanglement in fishing gear. Seabirds, marine turtles, large cetaceans, and small zooplankton often ingest plastic bits and trash like plastic bags, cigarette lighters, and bottle caps. Microplastics, are less than 5 mm and formed due to plastic embrittling in the presence of sunlight and seawater. Microplastics constitute a major chunk of oceanic plastic waste and can be consumed by zooplankton and other small marine animals. These microscopic plastic particles have been found in the organs of several aquatic species, including those that inhabit the deepest ocean trenches.

Chemical contaminants

Due to increasing industrialization, chemical waste levels have reached alarming heights in several parts of the world. The vast ocean expanse allows them to counteract certain chemical pollutants effectively. Acid rain can impact the tallest mountains as pollutants are deposited on their peaks and subsequently carried by water to areas where rivers flow into the ocean. Nevertheless, the accumulation of hazardous substances intensifies as more chemicals infiltrate the oceans. Except for human-caused activities, most chemical pollutants originate from factories, farms, and lawns. Chemical pollutants, specifically fertilizers, can enhance algal growth by utilizing photosynthesis, reducing the availability of oxygen and leading to the death of organisms that rely on oxygen for respiration. Many types of algae that form blooms produce toxins, which are subsequently concentrated and stored by organisms with limited detoxification capabilities.The concentration of these toxins increases significantly in organisms at higher trophic levels, leading to adverse effects and potentially even a population decline. Almost all non-biodegradable waste, including metals, can be classified within this category. Except for mercury, there appears to be no discernible correlation between environmental metal levels and the populations of fish and invertebrates. Bivalve molluscs can assimilate naturally occurring toxins from their surrounding environments due to their filter-feeding behaviour, similar to the microscopic organisms responsible for red tides. Research on the correlation between climate fluctuations and the presence of DDT and HCH in ice cores from Mt. Everest (in the Tibetan Plateau), Mt. Muztagata (eastern Pamirs), and the Rocky Mountains of North America indicates that mountain glaciers serve as repositories for DDT and hexachlorocyclohexane. The accumulation of DDT and highly chlorinated polychlorinated biphenyls (PCBs) in the food chain of both terrestrial and aquatic ecosystems has been documented in the Tibetan region. Wild fish species frequently contain recently identified pollutants, including polyfluoroalkyl and hexabromocyclodecanes.

Industrialization has increased chemical waste to dangerous levels worldwide. Oceans can neutralize chemical pollutants due to their size. As more chemicals enter the oceans, toxic materials accumulate. Most chemical pollutants come from factories, farms, and lawns (besides humans). Chemical pollutants, especially fertilizers, can boost algal productivity by using photosynthesis, limiting oxygen supply and killing oxygen-dependent organisms. Animals with weak detoxification systems accumulate toxins from bloom-forming algae.

Toxins as contaminants

Organisms may perish when exposed to specific toxins. A toxic substance is a chemical pollutant not indigenous to aquatic ecosystems. Specifically, herbicides, organochlorine pesticides, polychlorinated biphenyls, and chlorinated aliphatic hydrocarbons contribute to the elevation of water toxicity. Organic solvents, such as straight-chain surfactants, petroleum hydrocarbons, polynuclear aromatics, chlorinated dibenzodioxins, organometallic compounds, phenols, and formaldehyde, are known to contaminate water. Metals like lead, nickel, cadmium, zinc, copper, and mercury, as well as gases like chlorine, ammonia, and methane, and anions such as cyanides, fluorides, sulfides, and sulfites, in addition to acids and alkalies, are responsible for water pollution. Heavy metals inflict significant harm on living organisms across multiple levels. Urban industrial aerosols are the primary sources of heavy metal contamination.

The ocean is typically polluted by toxic substances that originate from maritime vessels. The origins of hazardous substances encompass municipal and industrial wastewater, runoff and leachate from solid waste disposal sites, industrial sites, storm sewer outfalls, urban areas, drainage from industrial sites, mines, and oil fields. The discharge from vessels, storage tanks, piles of chemicals, runoff from construction sites, sewer bypasses and sanitary pipes also contaminate the water. In addition, factors such as atmospheric deposition, highway runoff, logging operations, stormwater from urban areas and septic tank leachates, flow from agricultural fields and orchards, and forestry operations introduce pollutants into the aquatic environment.

When toxic substances are added to lakes, streams, rivers, seas, or other bodies of water, these substances undergo decomposition, flotation, sedimentation processes. Degradation of the accumulated substances results in water pollution, significantly impacting aquatic ecosystems. Moreover, pollutants can penetrate the underlying layers of soil and cause harmful effects on underground water reservoirs. Rivers are primarily polluted due to the substantial discharge of sewage and industrial waste. Presently, only a meagre 10 % of the wastewater generated undergoes treatment, while the remaining 90 % is discharged into water bodies without additional processing.

Consequently, contaminants infiltrate underground water, rivers, and other water reservoirs, potentially leading to the dissemination of waterborne diseases caused by pathogens in the water. Agricultural runoff, the water draining from cultivated land into rivers, substantially contributes to water pollution. The water may contain elevated levels of pesticides, nitrogen, and phosphorus.

Toxins accumulate in higher trophic levels, leading to adverse effects and potentially resulting in the collapse of populations. The condition encompasses the majority of waste that natural processes, such as metals, cannot break down. Except for mercury, there is no consistent correlation between environmental metal concentrations and the populations of fish and invertebrates. Filter-feeding bivalve molluscs can take natural toxins from their surrounding environments, like the microscopic organisms that cause red tide.

Research has observed that the presence of pulp mill effluent and other pollutants leads to increased fish coughing. Such a situation can harm fish's well-being and survival by significantly impeding their metabolic rate for migration or feeding and reducing their ability to spawn and produce offspring. Moreover, it could hinder fish growth and render them more susceptible to diseases and predators.

Microbes

Contaminants infiltrate groundwater, rivers, and other water bodies, and certain pollutants can lead to diseases caused by pathogens introduced into the water. Animal excrement often harbours pathogens that can lead to serious illnesses. Waterborne pathogens, such as bacteria, viruses, and protozoa, can lead to the transmission of diseases, ranging from mild respiratory and skin conditions to more severe illnesses like typhoid and dysentery. These microorganisms enter streams through untreated sewage, farm runoff, stormwater, and septic tanks. Moreover, there is a potential hazard of animal excrement, such as dead bodies and other animal by-products, polluting water sources due to inadequate disposal. Although small, these contaminants have a substantial effect, as demonstrated by their ability to induce illness. Bivalves can accumulate microbial contaminants such as sewage bacteria and viruses and chemical pollutants like metals and organochlorine compounds. The economies of numerous small coastal towns depend heavily on the molluscan shellfish industry. As filter feeders, shellfish accumulate sewage-derived bacteria and viruses in their tissues when exposed to sewage-filled streams. Consumers in numerous nations prefer partially cooked or raw shellfish, increasing the likelihood of acquiring illnesses such as cholera and typhoid from consuming infected tissue. Biologically impaired mussel beds considerably threaten public health and result in substantial economic losses.

The presence of microorganisms in drinking water in industrialized countries during the mid 1800s resulted in significant health issues in large urban areas. In the early 1900s, cities in Europe and North America initiated sewage infrastructure construction to redirect household waste away from water sources. Subsequently, these sewage and waste-treatment systems have experienced substantial growth in urban areas. Nevertheless, the ability of governments to enhance sewage and water infrastructure has not kept pace with the swiftly increasing urban population, especially in Asia and Latin America. In developed countries, waterborne infections have been eliminated. However, in developing nations, cholera outbreaks and other related diseases continue to be distressingly prevalent.

Conclusion

Contaminants or pollutants seriously threaten the aquatic ecosystem as they can adversely affect the ecosystem and its health. Aquatic water bodies, both freshwater and marine water bodies, are affected by this hazard. Therefore, monitoring aquatic health has to be carried out, closely monitoring water quality, quantity, and biological indicators. Thus, understanding the aquatic ecosystem and making informed decisions about sustainable water management are paramount. Monitoring activities involve collecting, analyzing, comprehending, and distributing information. Our ecosystem is facing a challenge sustain life anthropogenic and natural. The appropriateness of water for utilization, whether adequately pure or excessively contaminated, depends upon the measures implemented by the users. Aquatic ecosystems encounter multiple challenges, such as climate change, habitat degradation, chemicals, toxins, microbes, plastic pollution, disease, and modification of sediment loads. These problems have significant consequences for the inland and marine life. Therefore, to save life on Earth, contaminants must be reduced, if not eliminated, to fulfill sustainability concerning SDGs 2, 6, 13 and 14. Fish is the most important source for millions around the world the quality and quantity of water need to be conserved, which is the most important constituent of aquatic ecosystems.

Further Reading

Bashir, I., Lone, F.A., Bhat, R.A., Mir, S.A., Dar, Z.A. and Dar, S.A., 2020. Concerns and threats of contamination on aquatic ecosystems. Bioremediation and biotechnology: sustainable approaches to pollution degradation, pp.1-26

Gogoi, A., Mazumder, P., Tyagi, V.K., Chaminda, G.T., An, A.K. and Kumar, M., 2018. Occurrence and fate of emerging contaminants in water environment: a review. Groundwater for Sustainable Development, 6, pp.169-180

Pandey, P.K., Pande, A. 2024. Aquatic Environment Management 1st Edition ISBN: 9781003313137, p. 242

Rathi, B.S., Kumar, P.S. and Show, P.L., 2021. A review on effective removal of emerging contaminants from aquatic systems: Current trends and scope for further research. Journal of hazardous materials, 409, p.124413

13

Marine Fisheries Research in the Era of Global Cooperation Insights from the BOBP-IGO Model

P. Krishnan

Bay of Bengal Programme Inter-Governmental Organisation, Chennai-600 018 Tamil Nadu

Email: krishnanars@bobpigo.org

Introduction

We are living in at a time when we are all connected than ever – not only in prosperity, such as internet, but also in our dismays, such as unemployment, global food price inflation and climate change. On the global governance frontier, the 21st century has witnessed an unprecedented global commitment towards sustainable development, encapsulated through frameworks such as the Sustainable Development Goals (SDGs), the Biodiversity Beyond National Jurisdiction (BBNJ) Agreement, and the Global Biodiversity Framework (GBF). These frameworks signify a paradigm shift from the Millennium Development Goals (MDGs), which primarily focused on aid from developed countries to developing countries, towards a more inclusive and comprehensive approach to addressing global challenges. Unlike the MDGs, which had less emphasis on the means of implementation or the engagement of various stakeholders beyond governments and international organisations, the SDGs advocate for an inclusive approach to cooperation. Specifically, SDG 17, "Revitalise the global partnership for sustainable development," calls for a broad, multi-stakeholder partnership that encompasses governments, the private sector, civil society, the public, and international organisations. This recognition of the complexity of sustainable development challenges underscores the imperative for collective action and shared responsibility.

In the realm of marine biodiversity and fisheries management, the SDG-14, BBNJ Agreement and the GBF extend this ethos of global cooperation and shared responsibility to the conservation and sustainable use of marine biological diversity in the areas beyond national jurisdiction (ABNJ). These agreements highlight the critical need for international collaboration in addressing the challenges posed by overfishing, climate change, and habitat degradation in marine environments. They also emphasise the importance of preserving marine biodiversity as a global common good, necessitating a coordinated approach to governance and management that transcends national boundaries and interests.

However, despite the conducive framework of cooperation, there is growing concern about realized cooperation. Many people and Organizations would rather call the current era as the "Age of Distrust" manifested in what The Economist called the "Return of Economic Nationalism", wars and visible lack of progresses in global goals.

There is tremendous growth in academics and research in general and that in marine fisheries in particular. As of 2022, over 5.14 million academic articles are published each year, including short surveys, reviews, and conference proceedings. Notably, over 90 % of these publications came from countries with high-income and upper middle-income economies. In case of fisheries sector also, there is likely an exponential growth of research as measured by publications. The largest contributors to fisheries research include the USA, China, Japan, Australia, Canada, and Norway. India is an emerging player from the Bay of Bengal region.

Set in this context, the paper explores the barriers and opportunities in promoting regional research and regional collaboration in development. The Bay of Bengal (BOB) region, characterised by its diverse fisheries and reliance on shared stocks, serves as a microcosm of the broader challenges and opportunities presented by these global development frameworks. It has necessitated a nuanced understanding of the interplay between global development frameworks and local realities, highlighting the importance of aligning fisheries research and management strategies with the principles of sustainability, inclusivity, and cooperation outlined in the SDGs, BBNJ Agreement, and Global Biodiversity Framework.

Barriers and opportunities for collaborative research

Collaborative marine fisheries research faces several challenges that can also offer valuable lessons for improving practices and policies. Here are some key lessons that can be drawn:

- ***Language and Communication Barriers:*** The dominance of English as a working language can marginalise non-English speakers. This barrier highlights the need for multilingual collaboration and translation services to promote more inclusive communication in research efforts.

- ***Exclusive Agenda Setting:*** Powerful stakeholders often shape research agendas, creating a barrier to inclusive dialogue. Opportunities arise from this challenge when efforts are made to democratise agenda-setting and ensure stakeholder representation from all sectors, including the global south.
- ***Pre-existing Power Dynamics:*** Acknowledging the influence of existing power relationships is crucial. This challenge presents an opportunity to build the capacity of marginalised stakeholders and advocate for equitable representation in research collaborations.
- ***Inclusivity and Equity:*** The need for inclusive participation is paramount for achieving equitable research outcomes. Lessons can be learned by actively involving a broad spectrum of states, local communities, and indigenous peoples in research projects.
- ***Socioeconomic Disparities:*** There may be disparities in economic and scientific capacities among different countries and communities. Collaborative efforts must strive to provide support and resources to balance these disparities.
- ***Regulatory Constraints:*** Researchers must navigate a variety of laws and regulations, which can be a barrier to collaborative work. However, these stipulations also bring an opportunity to establish shared protocols and ethical standards that support responsible research activities.
- ***Challenges of Scale:*** Multiscale dynamics require research approaches that are flexible and adaptable. By recognising the interplay between local, regional, and international factors, collaborative research can better address the multifaceted challenges facing marine fisheries.
- ***Ownership and funding:*** The national research programmes and funding mechanisms do not outline the need for collaborative research. This is another type of regulatory constraint. However, even when approaches are made to remove the regulatory barriers, e.g. through participation in a regional platform, organizations usually lack a funding mechanism for activities beyond national jurisdiction. Further, effective coordination among multiple international partners requires significant effort, with logistical challenges often impeding research progress.
- ***Data Sharing and Intellectual Property:*** Concerns over data ownership, privacy, and intellectual property rights can limit the willingness of entities to share crucial research data.

On the other hand, collaborative research in marine fisheries presents a wide range of opportunities that can significantly contribute to the advancement of the blue economy and address global concerns such as overfishing and fleet overcapacity:

- ***Enhanced Sustainability:*** Through collaborative efforts, research can provide a comprehensive understanding of fish stock health and sustainable harvesting levels, helping to prevent overfishing and support the regeneration of marine ecosystems.
- ***Shared Knowledge and Resources:*** Different countries and organisations can pool their expertise, data, and financial resources to conduct more extensive and impactful research, leading to better-informed management practices.
- ***Innovative Technologies and Methodologies:*** Collaboration offers the chance to develop and apply new technologies, from satellite monitoring to advanced modelling techniques, for improved fisheries management and enforcement against illegal practices.
- ***Adaptive Management Strategies:*** Collaborative research can inform adaptive management strategies that respond to dynamic ocean conditions and human impacts, ensuring the resilience of marine ecosystems and fisheries livelihoods.
- ***Policy Harmonisation:*** Shared research findings can facilitate the harmonisation of policies and regulations across jurisdictions, which is crucial to tackling overfishing and fleet overcapacity effectively, especially in shared waters and migratory fish stocks.
- ***Capacity Building:*** Collaboration presents an opportunity to build capacity in developing countries, enabling them to participate in research and management and to benefit economically from their marine resources under the blue economy framework.
- ***Ecosystem-based Management Approaches:*** By working together, researchers can develop and implement ecosystem-based management plans that integrate fisheries with other sectors, such as tourism and marine transport, optimising resource use within the blue economy.
- ***Science-Policy Interface:*** Collaborative research can serve as a bridge between scientists and policymakers, ensuring that research findings directly inform policy and regulatory decisions, leading to effective and timely responses to issues like overfishing and fleet overcapacity.

By addressing issues like overfishing and fleet overcapacity through a collaborative approach to research, the full potential of marine fisheries within the blue economy can be realised, providing economic benefits while ensuring the sustainability and health of marine ecosystems.

The South-South Model of Collaboration in Marine Fisheries Research

The South-South cooperation model (SSCM) involves the partnership between developing nations (also referred to as countries of the Global South) to pursue joint development objectives, and further advance economic growth and social advancement. This approach serves as an alternative to traditional North-South collaboration, where support and cooperation originate from developed countries for developing ones.

SSC operates on the basis of solidarity and mutual benefit among participating nations, emphasising equality, respect for sovereignty, and the exchange of resources, technology, and knowledge among Global South countries. The underlying concept is that these nations face similar challenges and can, therefore, potentially offer more pertinent and sustainable solutions by drawing upon their own experiences than those provided by developed countries.

This type of cooperation can involve various forms, including trade agreements, technology transfers, sharing of best practices, joint ventures, and capacity-building projects. The cooperation is not only limited only to economic aspects but also extends to political, social, and cultural exchanges. The goal is to promote self-reliance among developing countries by tapping into their potential and fostering partnerships that address their specific needs and priorities (Amanor and Chichava, 2016). In the context of marine fisheries research, the South-South model of cooperation can play a significant role in addressing issues related to resource management and sustainability (Kao and Pearre, 2018).

In the research field of fisheries science, a recent publication (Syed *et al.*, 2019) reported that the prevalent structure of international collaborations largely mirrors the north-south model where western economies are the clear leaders. This pattern has the potential to intensify the extant disparities in the dissemination and impact of scientific knowledge. They noted that the scarcity of South-South collaborations marks a critical concern, as these partnerships are essential for fostering scientific self-sufficiency and meeting the unique demands of developing countries.

This north-south divide is further accentuated by the visibility and citation. Developed countries tend to exhibit an international orientation in their research endeavours, resulting in heightened visibility and citation frequency due to their presence in globally recognised journals. On the contrary, emerging economies find their research primarily enclosed within national publications, which curtails their global presence and influence, reinforcing the marginalisation of these voices in the scientific discourse.

Analysis reveals that the currents of international scientific collaboration in fisheries science exhibit a pronounced regionalised configuration rather than a globally integrated network. Factors such as geographic proximity, along with political, economic, historical, and cultural ties, continue to significantly influence the formation of research collaborations, speaking to a preference for partnering within more familiar or accessible confines (Syed *et al.*, 2019).

The Impact of Global Science Divide in Fisheries Research

The immediate and direct impact of the global science divide is the loss of voice of the global south in the development narrative and global agenda setting. Experiences from global platforms such as FAO Committee on Fisheries (COFI), World Trade Organization (WTO) to regional fisheries management organizations (RFMOs) showed that while western countries present a coordinated voice backed by scientific evidence, the global south remain isolated in their individual voices due to gaps in scientific backstopping. While in the recent past Africa and Latin America have presented a unified stand, it was limited to political objectives as the scientific inputs still largely come from the western countries.

Given these insights, it is imperative that future strategies in capacity building, particularly within the Global South, prioritise not only quantitative increases in collaborative efforts but also a qualitative shift towards embracing a plurality of scientific narratives. By supporting a more equitable research landscape, the scientific community can endeavour to democratise knowledge production and inclusion, ensuring that the wealth of diverse perspectives is recognised and valued on the global stage (Syed *et al.*, 2019). This recommendation aligns with the broader sentiments expressed by the United Nations' Sustainable Development Goals and the Global Oceans Science Report, which advocate for enhancing the research output from underrepresented regions to address the inequalities between the Global North and South (Syed *et al.*, 2019).

The BOBP-IGO Model

Established in 2003 as a regional fisheries body, the Bay of Bengal Programme Inter-Governmental Organisation is a shining example of SSC. The forerunner of BOBP-IGO was the Bay of Bengal Programme of FAO, a field project that was largely funded by multilateral donors (a North-South model). However, as the project matured, the project countries decided to internalise it and entered into the BOBP-IGO Agreement, making it the first RFB that is wholly funded and run by developing countries.

The inspiration behind the setting up of the BOBP-IGO was to accentuate collaboration in various aspects of the marine fisheries. The Technical Advisory

Committee of the BOBP-IGO, comprising leading fisheries research agencies of its member countries, Bangladesh, India, Maldives and Sri Lanka were a platform for exchanging ideas and developing collaborative programmes. The successful engagement led to various key developments in the region including a collaborative research programme on hilsa (Bangladesh, India and Myanmar), sharks (Bangladesh, India, Maldives and Sri Lanka), safety at sea for small-scale fisheries (Bangladesh, India, Maldives and Sri Lanka) and fisheries monitoring, control and surveillance and subsequent developments of draft regional plans of action on shark and illegal, unreported and unregulated fishing.

However, while these initiatives are continuing, they cannot be scaled up, and the momentum cannot be sustained as it proved challenging to put these initiatives on autopilot. Project-based funding of these activities was also not conducive to long-term sustainability.

In the wake of valuable lessons learned from past initiatives, the BOBP-IGO has embarked on innovative endeavours such as BOB-SAN (Bay of Bengal Stock Assessment Network) and BIMReN (BIMSTEC-India Marine Research Network). These initiatives underscore a commitment to cost-efficient cooperation models, harnessing the power of modern communication technology and aligning with the emerging global framework. Additionally, they align with India's evolving global outlook, particularly exemplified by the Neighbourhood First Policy also has created, creating a conducive environment for fostering regional cooperation in marine research. This strategic shift recognises the need for collaborative efforts to address the complexities of fisheries management in the Bay of Bengal region.

One such emerging framework is BIMSTEC-India Marine Research Network (BIMReN), an initiative supported by Ministry of External Affairs, Government of India and implemented by BOBP-IGO to facilitate close network among the researchers from Bay of Bengal region. BIMReN provides scope for the citizens of BIMSTEC countries to pursue doctoral research in Indian Universities and Indian Institutions and to undertake joint research projects with other countries.

***Bay of Bengal Stock Assessment Network**:* BOB-SAN, functioning as an informal virtual network with BOBP-IGO as its nodal point, aims to leverage the collective wisdom and experiences of its members to enhance stock assessment methodologies and contribute to coordinated regional fisheries management. Understanding the intricacies of fish stock assessment in the Bay of Bengal region, BOB-SAN has committed to meeting periodically, convening once every six months. These gatherings serve as forums for sharing knowledge and best practices in fish stock assessment, discussing advanced methodologies, and collectively advancing the frontiers of stock assessment based on cumulative experiences and collaboration. The network operates on the premise that addressing the complexities of fisheries management in the Bay of Bengal requires a unified effort, with each member contributing to the shared goal of sustainable and effective stock assessment in the region.

BIMSTEC-India Marine Research Network (BIMReN) would serve as a platform for regional collaboration in research on regional development issues, fostering BE in the Bay of Bengal Region. BIMReN is in furtherance to the main objectives of BIMSTEC, namely improving people-to-people contact and cultural cooperation and contributing to other BIMSTEC objectives, such as promoting fisheries, aquaculture, trade, and energy. BIMReN includes three principal components: BIMReN Split-Site Doctoral Fellowship, BIMReN Twinning Research Project and BIMReN Blue Economy Conference.

Conclusion

The path towards sustainable fisheries management is complex and requires a multifaceted approach that encompasses technological innovation, inclusive governance, capacity building, and international cooperation. By adhering to the principles laid out in global frameworks such as the SDGs, SSC, the BBNJ Agreement, and the Global Biodiversity Framework, we can navigate the challenges of sustainable resource use. The strategic pathways outlined above offer a blueprint for transforming the fisheries sector into a model of sustainability, resilience, and shared prosperity, ensuring the conservation and sustainable use of marine resources for future generations.

References

Amanor, Kojo, and Sérgio Chichava. "South–South Cooperation, Agribusiness, and African Agricultural Development: Brazil and China in Ghana and Mozambique." Elsevier BV, vol. 81, 1 May. 2016, p. 13-23. https://doi.org/10.1016/j.worlddev.2015.11.021.

Haas, Bianca, *et al.* "The use of influential power in ocean governance." Frontiers in Marine Science 10 (2023): 977.

Kao, Shih-Ming, and Nathaniel S. Pearre. Regional Cooperation in the Post-South China Sea Arbitration Era: Potential Mechanism and Cooperative Areas. 8 Feb. 2018, https://doi.org/10.1080/08920753.2018.1426377.

Syed, Shaheen, *et al.* "Mapping the global network of fisheries science collaboration." Fish and Fisheries 20.5 (2019): 830-856.

14

High Seas Treaty and Its Relevance to Marine Fisheries in India

E. Vivekanandan

Senior Scientific Consultant, Bay of Bengal Programme Inter-Governmental Organisation, Chennai - 600018, Tamil Nadu

Email: evivekanandan@hotmail.com

Introduction

The United Nations High Seas Treaty, also known as the Biodiversity Beyond National Jurisdiction treaty or the BBNJ treaty, is a legally binding instrument for the conservation and sustainable use of marine biological diversity in areas beyond national jurisdiction. It is an agreement under the United Nations Convention on the Law of the Sea (UNCLOS). The text of the Treaty was finalised during an intergovernmental conference at the UN on 4 March 2023 and adopted on 19 June 2023. States and regional economic integration organizations can become parties to the treaty.

The treaty addresses four themes: (1) Marine genetic resources, including fair and equitable sharing of benefits; (2) Area-based management tools, including marine protected areas (MPAs); (3) Environmental impact assessments (EIAs); and (iv) Capacity building and transfer of marine technology (Figure 1). The area-based management tools and environmental impact assessments relate mainly to conservation and sustainable use of marine biodiversity, while the marine genetic resources and capacity building and transfer of marine technology include issues of economic justice and equity.The agreement provides for the common governance of about half of the Earth's surface and 95% of the ocean's volume, the largest habitat on our blue planet, to promote equity and fairness, tackle environmental degradation, fight climate change, and prevent biodiversity loss in the high seas.

The Treaty opened to State signatures on 20 September 2023. By signing, countries express their willingness to proceed to ratification, when they formally consent to the new international law. As soon as 60 countries of the total 193 countries sign and ratify, the Treaty will enter into force and international action can be taken up to conserve to protect the shared ocean, mitigate climate breakdown and safeguard the lives and livelihoods of billions of people worldwide. As on date, 84 countries have signed the treaty, but the treaty is yet to be ratified by any country. India has not signed and ratified the Treaty so far.

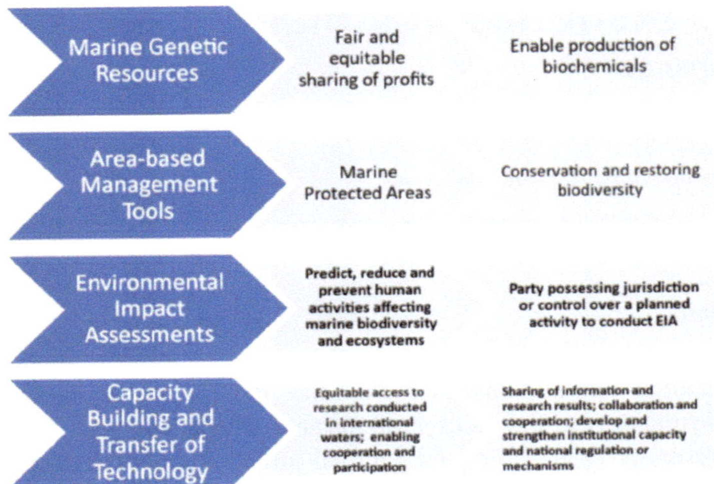

Fig. 14.1: The content of the Treaty

Relevance to marine fisheries

Fishing is one of the major economic activities in the high seas. High sea fish species are caught by industrial fleets and destined mainly for high-end markets. More than 80% of high sea fishing is done by five countries, namely China, Chinese Taipei, Japan, South Korea and Spain. All these countries have signed the Treaty.The number of boats engaged in high seas fishing is reported as 3620 (Global Fishing Watch, 2018; Carmine *et al.*, 2020). One-third of the fishery is operated by about 100 corporate companies. While 6% of total global marine fish catch is contributed by high sea fishing with a value of 7.6 billion US$ (Table 1), fishing at the current scale is enabled by large government subsidies, without which as much as 54% of the present high-seas fishing grounds would be unprofitable at current fishing rates (Sala *et al.*, 2018). The patterns of fishing profitability vary widely between countries, types of fishing, and distance to port.

Table 14.1: Global high sea fishing fleet and economics (Source: Global Fishing Watch, 2018)

Parameters	
No. of vessels	~3620
Catch (million t)	4.4 (6% of total catch)
Value (billion $)	7.6 (8%)
Cost (billion $)	6.2 – 8.0
Profit (billion $)	-0.3 – 1.4
Subsidy (billion $)	4.2
Profit (billion $), with subsidy	3.8 – 5.6

Large, mobile, pelagic, high-value fishes (tunas & sharks) are targeted (about 99% of catch) by longliners, gillnetters, purse-seiners. All other fishing methods like the bottom trawling and squid jiggers are non-profitable. Though the stock status of the target species has not been estimated properly, the available estimates indicate a decline in the abundance of many high sea species. For a proper understanding of the stock status of these fishes, high sea fish biodiversity and stock biomass need to be estimated adequately. Moreover, fishing impacts a wide range of species beyond those targeted (like dolphins, turtles, sea birds), which should be taken into account.

During the UN negotiations it was a contentious point whether or not marine genetic resources should apply to fish and fishing activities. At present, 17 Regional Fisheries Management Organisations (RFMOs), have management roles of high seas fisheries. The RFMOs set catch limits and bycatch mitigation measures for high seas fisheries for select species such as tunas, jack mackerel, and swordfish in different regions of the ocean. Considering this, the final High Seas Treaty states that the provisions about marine genetic resources do not apply to 'fish' and 'fishing' in areas beyond national jurisdiction. This decision is likely to impact the ability of the Treaty to address its objective, since fish are a major component of marine biodiversity and play an essential role in the functioning of marine ecosystems. It should be noted that although most areas beyond national jurisdiction fall under at least one RFMO, their mandates are narrowly focused on managing a limited number of commercially important fish species and associated bycatch. In a recent assessment, it has been found that the organizations have made only moderate progress in managing bycatch species such as sharks, seabirds, marine mammals and turtles incidentally caught in these fisheries. The RFMOs have made only limited progress in taking a more holistic approach to the impact on the food web, habitats, and broader ecosystems. Considering this, it is critical for the High Seas Treaty to embrace an integrated and comprehensive view of biodiversity, including fish biodiversity, so that the efforts of RFMOs and the Treaty concerning protecting biodiversity beyond national jurisdiction complement one another. Establishing large MPAs outside national jurisdiction may help protect high sea fishes. However, the question remains as how MPAs would be protected from commercial fisheries and prevented from Illegal, Unreported and Unregulated fishing in the high seas.

Relevance to marine fisheries in India

Marine fisheries in India are predominantly small-scale in nature, operating mostly in coastal waters, restricted within the EEZ. At present, there is no organised high sea fishing. However, about 900 boats occasionally fish in the high seas. These boats are operated by the fishermen from Kanyakumari district using medium-sized fibre-board/steel/wooden boats. They are highly skilled in fishing with indigenous knowledge, targeting tunas and sharks without employing modern technologies like powered winch, on-board fish preservation and processing facilities. There is no accurate data on their fishing activities like the fishing grounds, number of fish catch, fishing duration and species composition. Often, the fishermen travel through the high seas and fish in other countries within their EEZ and are caught and arrested.

Government of India made significant efforts in the last 40 years to introduce capital-intensive fishing techniques with foreign assistance in technology and expertise such as Chartering fishing vessels (1981), Deep Sea Fishing Policy (1991), Joint Venture, import foreign vessels under Letter of Permit (LoP) of EXIM policy (2002). However, these measures were severely resisted by fishers because the communities, to a large extent, were not benefited. Consequently, these schemes did not succeed and were discontinued midway. Expert committees were constituted time and again to suggest measures to promote high sea fisheries. Fishers demanded government support to modernise their fishing fleet. Following this, the GoI made efforts to meet the demands of the fishers. (i) In 2017, the government announced measures by financing fishermen under Blue Revolution. (ii) Under PMMSY, the GoI supported Acquisition of Deep-Sea Fishing vessels by traditional fishermen. (iii) Under Centrally Sponsored Scheme, the GoI is implementing a scheme "Conversion of Trawlers into Resource Specific Deep-Sea Fishing Vessels". (iv) In a Tamil Nadu-specific scheme "Diversification of Trawl Fishing Boats from Palk Straits into Deep Sea Fishing Boats (longliners and gillnetters)" aimed at providing 2,000 vessels to fishermen of the State and motivating them to abandon bottom trawling in the Palk Bay. (v) Under Deep Sea Mission, bio-prospecting of deep-sea flora and fauna including microbes and studies on sustainable utilisation of deep-sea bio-resources are the main focus.

GoI also prepared draft guideline on high sea fishing. Under the guideline, no person shall engage in any fishing operation in the high seas except under the authority of a valid permit granted by the Issuing Authority. Boat markings and communication requirements will be made following FAO standards. Logbook should be maintained, and details of fishing should be declared after every voyage. Standard safety protocol has to be maintained. Only selected types of gears will be permitted to be operated. Without prior permission, trans shipment at sea and unloading catch are prohibited. The fishers also cannot catch, trade or possess scheduled species.

Reforms needed to promote and sustain high seas fishing

For developing sustainable high sea fishing, India needs to develop a strong and viable reform (Fig. 2). In the context of High Seas Treaty, being a coastal fishing nation, India has to consider a few critical questions: What should be the scale of investment in high sea fishing? Will high seas fishing be economically viable and sustainable? Should India support large MPAs in the high seas? Will High Sea MPAs improve coastal fishing considering the inter-connectedness of the seas ('spillover' effect)?

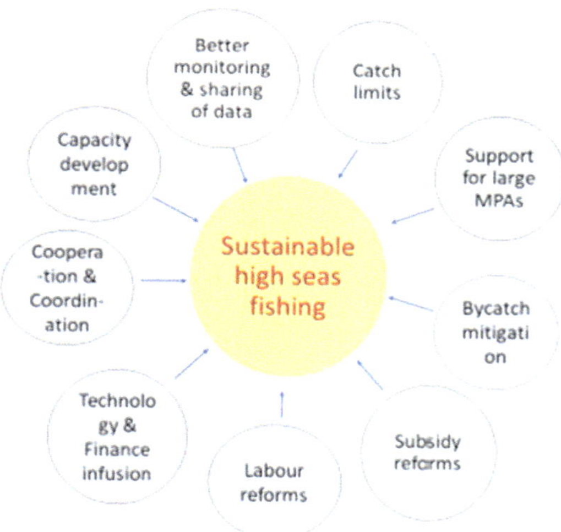

Fig. 14.2: Reform needed for sustainable high seas fishing by India

India's potential role

Although the High Seas Treaty does not directly address fisheries, and India's high seas fishing is not fully developed, the country may consider signing and ratifying the Treaty. Ratifying the High Seas Treaty aligns with India's commitments to SDGs like Climate Action and Life Below Water. It presents a unique opportunity to protect our seas and combat the challenges of climate change, biodiversity loss and pollution. India, as a responsible global player, should use this opportunity by becoming a signatory to the Treaty.

References

Carmine, G., J. Mayorga, N.A. Miller *et al.*, 2020. Who is the high seas fishing industry? One Earth, 3: 730-738.

Global Fishing Watch, 2018. The year in transparency. www.global fishing watch.org.

Sala, E., J. Mayorga, Costello, *et al.*, 2018. The economics of fishing the high seas. Science Advances, 4: eaat2504.

15

Recent Advances in Socio-economics of Inland Fisheries in India

***Pradeep K. Katiha*[1] *and Anjana Ekka*[2]**

[1]*Plan Implementation and Monitoring Section, Indian Council of Agricultural Research, New Delhi*

[2]*ICAR-Central Inland Fisheries Research Institute, Barrackpore-700 120 West Bengal*

Email: pkatiha60@gmail.com

Introduction

Inland fisheries in India represent a dynamic sector that plays a pivotal role in the country's fisheries and aquaculture landscape. India stands as the world's third-largest fish producer and the second-largest in aquaculture. The Blue Revolution in India highlights the significance of the fisheries and aquaculture sector, positioning it as a crucial player in national economy. In past, there has been a notable shift from marine-dominated fisheries to inland fisheries, with latter now contributing ~70% of fish production, up from 36% in the mid-1980s. This transition has paved the way for a sustainable blue economy.

The extensive network of rivers, including 14 major and 44 medium rivers provides source of original germplasm of native fish species which needs to be conserved towards serving the aquaculture sector for quality seed and better growth. Conservation and awareness in riverine fisheries programs, under the Blue Revolution, focus on indigenous fisheries resource conservation and restoration of natural productivity through river ranching. It is important for establishment of a sustainable blue economy in the country. Despite substantial growth in inland fisheries and aquaculture, the sector has yet to realize its full potential. Vast untapped resources, particularly 1.2 million ha of floodplain lakes, 2.36 million ha of ponds and tanks, 3.54 million ha of reservoirs, and 1.24 million ha of brackish water resources, offer significant opportunities for increased production, livelihood development, and economic prosperity. Reservoirs, often termed 'sleeping giants,' contribute only ~3.81% to total inland fish production through cage culture. The PMMSY aims to unlock the potential of 3.54 million ha of reservoirs by promoting culture-based fisheries and sustainable cage culture, targeting a production increase from 2.44 million t to 6.29 million t by year 2024-25.

Inland fisheries are intricately tied to socio-economic factors, playing a pivotal role in the lives of communities residing near freshwater bodies like lakes, reservoirs and rivers. One of the primary dimensions where socio-economics comes into play is in providing livelihoods and employment opportunities. Fishing activities directly engage fishers and those involved in processing, transportation, and marketing, thereby contributing to local economies. The impact extends beyond, supporting the livelihoods of numerous individuals in related industries. Culturally and socially, inland fisheries hold immense significance. Traditional fishing practices, rituals, and community gatherings are influenced by socio-economic factors. Moreover, socio-economic factors influence the ability of communities to adapt to the impacts of climate change on inland fisheries. Vulnerable communities may face challenges in adapting to changing fish populations and ecosystems, necessitating integrated approaches that consider both social and economic dimensions.

In this regard, addressing and understanding socio-economic factors are imperative for the sustainable management of inland fisheries. Policies and interventions should consider the intricate relationships between human communities, economic activities, and the ecological health of freshwater ecosystems, ensuring the well-being of both fisheries and the communities they support over the long term. Therefore, the chapter describes i) Social aspects of inland fisheries in Indian communities, ii) Socio-economic researchable and monitorable issues in inland fisheries, iii) Economic contribution of inland fisheries in India, iv) Policies and measures for development of inland fisheries in India, v) Challenges and opportunities in India's inland fisheries sector, vi) Sustainable practices in inland fisheries management and vii) Future directions for inland fisheries in India.

Social aspects of inland fisheries in Indian communities

Inland fisheries in India exert a profound influence on the social fabric of communities, encompassing a spectrum of aspects that shape livelihoods, cultural practices, and community dynamics. The social dimensions of inland fisheries are intricately interwoven with the daily lives of individuals, and understanding these aspects is crucial for a comprehensive view of the sector's impact on society. Inland fisheries are a major source of livelihood for a significant portion of the population, especially in rural areas. According to data, the sector provides employment to over 14 million people, both directly and indirectly. Traditional fishing communities, often belonging to marginalized sections, depend on inland fisheries for sustenance. Their social identity and cohesion are closely tied to their shared dependence on fishing-related activities. Fishing activities have deep-rooted cultural significance in many Indian communities, contributing to their traditions and social practices. Fishing festivals, rituals, and ceremonies are common in regions where fishing is a traditional occupation. The knowledge and skills associated with fishing techniques are often passed down through generations, fostering a sense of cultural continuity. The establishment of infrastructure related to inland fisheries, such as landing centers, storage facilities, roads, community halls, *etc.* contributes to the development of social infrastructure in fishing communities. These facilities enhance the overall quality of life. The social aspects of inland fisheries in Indian communities are diverse and deeply embedded in the cultural and economic fabric of society. As the sector continues to evolve, sustainable practices and community-based approaches will play a crucial role in maintaining the social harmony and well-being of these communities.

Socio-economic researchable and monitorable issues in inland fisheries

The research and monitoring of socio-economic issues are important for understanding the status of any community with respect to their economic, social, cultural, political, health and physical assets scenario. The indepth study of these issues would be useful in preparation of roadmap for the development of the community. Therefore, the researchable issues and the indicators are summarized (based on Bennet *et al.*, 2021) below:

Economic issues

The economic issues are related to fishers' livelihood, employment and financial activities. The research and monitoring attributes are a) Employment, b) Income and wage structure; c) Poverty index; d) Livelihood diversification and security covering alternative livelihood options, e) Market environment including freedom to sell and remuneration f) Financial performance, *i.e.* income over cost/investment, g) Economic viability and h) Contribution of sector to national income.

Social issues

The social issues entail the social relationships, organizations and supports in the society. The researchable/monitorable issues may encompass: a) Gender issues including women participation in fisheries, b) Equity conflict regarding sharing of economic benefits and conflict and/or competition over the resource, c) Community relations and social capital i.e. Sense of belonging in community, d) Organizational supports, e) Education and training and f) Social Resilience or Adaptive Capacity including access to assets.

Cultural issues

The cultural aspects include the status of cultural wealth, institutions, customs, *etc.* including a) Traditional knowledge and access to knowledgeable elder fishers, b) Cultural practices and activities c) Heritage including interactions with environment over generations, d) Access to areas needed for cultural activities, and e) Cultural entity of being connected to fisheries.

Political issues

The political empowerment of any community may be assessed by their inclusiveness, quality and fairness of governance and decision-making processes. The researchable attributes may be a Participation and voice in management decisions, b) Transparency and access to information, c) Accountability, d) Rule of law, e) fairness in allocation of harvesting or area rights for fishing, f) Legitimacy of fisheries regulations and management, g) Resource Access and tenure, h) Access to justice and conflict resolution mechanism, i) Presence of legal recognitions of right to self-determination

Health issues

The health issues cover physical, mental and psychological health condition of fishers including occupational and harvesting safety and perceptions of life satisfaction. This also encompass aspects related to food and nutritional security.

Physical assets and infrastructure

These issues pertains to the physical assets of individuals and communities that support economic activities and other aspects of their well-being. The information may be collected on assets and community infrastructure for a) Fishing, b) Schools and daycares, c) Processing, d) Crafts and gears, *etc.*

Economic contribution of Inland Fisheries in India

The economic contribution of inland fisheries in India is profound and multifaceted to the nation's economy through various channels. As the sector has evolved and expanded, its influence on income generation, employment, food security, and rural development has become increasingly pronounced. As the sector continues to evolve and modernize, its role as a driver of economic growth and social development is expected to become even more prominent.

Fig. 15.1: Socio-economic contribution of fisheries in Inland sector

Inland fisheries provide a crucial source of income for millions of people, especially those residing in rural areas. Fishers, fish farmers, and those engaged in related activities such as fish processing and marketing contribute substantially to household incomes. The sector's labour-intensive nature contributes to the creation of both direct and indirect employment, fostering economic growth in rural and peri-urban areas. Inland fisheries play a crucial role in ensuring food security by providing a protein-rich source of nutrition to a large population. Fish is a staple food for many communities, and the availability of diverse fish species from inland waters enhances dietary diversity. The establishment of fishery-related infrastructure, such as landing centers, hatcheries, and processing units, further enhances the socio-economic fabric of rural regions and thus inland fisheries contribute to the overall development of rural areas by providing a sustainable source of income, which, in turn, stimulates economic activities in local communities.

Inland fisheries also contribute to India's export earnings. The economic impact is evident in the global market, where Indian fish and seafood products find demand, boosting foreign exchange reserves. Inland fisheries also provide ecological services by contributing to the maintenance of aquatic ecosystems.

Well-managed or responsible fisheries practices contribute to biodiversity conservation, water quality improvement, and ecosystem stability. Inland fisheries often serve as attractive destinations for recreational fishing and tourism, contributing to local economies. This creates additional economic opportunities through hospitality services and tourism-related businesses.

Policy and measures for the development of inland fisheries in India

India has implemented several policies and measures aimed at regulating, promoting, and sustaining the inland fisheries sector, recognizing its significant economic, social, and environmental contributions. These policies encompass various aspects such as resource management, conservation, technology infusion, and community engagement. Some of the key policies and measures in inland fisheries in India are enlisted below:

National Fisheries Policy

The National Fisheries Policy is to harness the full potential of the fisheries sector, including inland fisheries, by promoting sustainable practices, enhancing productivity, and ensuring the welfare of fishing communities.

Prime Minister's Matsya Sampada Yojana (PMMSY)

PMMSY was launched to focus on sustainable development, increasing fish production, enhancing livelihoods, and boosting the overall economy during the financial year 2020-21 to FY 2024-25 for holistic development of the fisheries sector in the country. So far, under PMMSY investments worth Rs. 4005.96 crore have been approved during the last three financial years (FY 2020-21 to 2022-23), and current financial year (2023-24).

River Ranching Program

Under the Blue Revolution, the River Ranching Program focuses on the conservation of indigenous fisheries resources and the restoration of natural productivity in rivers. This program involves the stocking of native species to promote sustainable riverine fisheries.

Cage Culture Promotion in Reservoirs

To unlock the vast potential of reservoirs and floodplain wetlands, the government has initiated policies promoting cage culture. The goal is to optimize production through culture-based fisheries, emphasizing the sustainable use of these water bodies.

Community-Based Management

The government policies encourage community-based management systems, empowering local communities in the decision-making processes related to fisheries. This approach enhances social cohesion, ensures sustainable resource use, and promotes responsible fishing practices.

Conservation of Biodiversity

Policies emphasize the conservation of biodiversity in inland water bodies. Efforts include the protection of native fish species, implementation of regulations against destructive fishing practices, and the restoration of degraded habitats.

Technology Infusion and Capacity Building

The government invests in technology infusion and capacity building programs to equip fishers with modern techniques, ensuring sustainable practices. This includes training programs on improved aquaculture methods, disease management, and the use of advanced equipment.

Promotion of Women in Fisheries

Policies recognize the significant role of women in the fisheries sector, especially in post-harvest activities. Initiatives focus on empowering women through training, financial assistance, and promoting their active participation in the decision-making processes. These policy measures collectively contribute to the holistic development and sustainability of inland fisheries in India. By integrating environmental conservation, community involvement, and technological advancements, these policies aim to propel the sector towards greater productivity, economic growth, and social well-being.

Challenges and Opportunities in India's Inland Fisheries Sector

The inland fisheries sector of India navigates a dynamic landscape marked by both challenges and opportunities. As the sector undergoes transformations driven by factors such as changing environmental conditions, technological advancements, and evolving socio-economic dynamics, addressing these challenges and leveraging opportunities becomes imperative for sustainable development (Katiha *et al.*, 2017; Ekka *et al.*, 2012).

Challenges in the sector include the persistent threats of overexploitation and habitat degradation. Unregulated fishing practices, pollution, and habitat destruction can lead to declining fish stocks and ecosystem degradation. Additionally, climate change introduces uncertainties with altered precipitation patterns, rising temperatures, and extreme weather events, impacting the distribution and abundance of fish species, thus affecting the productivity of inland water bodies.

Resource use conflicts also pose a significant challenge, particularly in densely populated regions. Competing interests for water resources and land use can lead to tensions between agriculture, industry, and fisheries. Inadequate infrastructure, such as storage facilities, transportation networks, and market linkages, further hinders efficient post-harvest management and marketing of fish produce, limiting economic opportunities for fishermen.

On the flip side, there are several opportunities that can be harnessed for the sector's growth. The integration of modern technologies, including aquaculture practices, precision farming techniques, and digital tools, presents opportunities for improving productivity and reducing environmental impact. Sustainable aquaculture practices, such as cage culture and integrated farming systems, offer avenues for increased fish production while minimizing environmental impacts.

Government initiatives, such as the Prime Minister's Matsya Sampada Yojana (PMMSY), allocate funds for the development of inland fisheries. These programs focus on enhancing infrastructure, capacity building, and conservation efforts, offering a comprehensive approach to sectoral growth. Community-based management and empowerment programs can foster sustainable practices, enhance local livelihoods, and promote responsible resource use.

Diversifying fish species through the introduction of high-value and resilient species broadens market opportunities and improves the economic resilience of the sector. Conservation efforts, including river ranching programs and biodiversity protection measures, contribute to the sustainable management of inland fisheries. In conclusion, while challenges persist, the strategic pursuit of opportunities can pave the way for a resilient, economically vibrant, and ecologically sustainable future for India's inland fisheries sector.

Sustainable Practices in Inland Fisheries Management

Sustainable management of inland fisheries is contingent upon the adoption of practices that harmonize ecological well-being, economic viability, and social welfare. These practices are indispensable for ensuring the enduring resilience and productivity of inland fisheries. Key components of sustainable inland fisheries management encompass an ecosystem-based approach, robust regulatory frameworks, community-based management, stock assessment, habitat conservation, selective fishing gear usage, integrated aquaculture practices, education and capacity building, climate resilience, and certification initiatives.

Implementing an ecosystem-based approach involves recognizing and understanding the intricate connections within the ecosystem, prioritizing habitat preservation, and mitigating anthropogenic impacts on biodiversity.

Establishing and enforcing effective regulatory frameworks, including guidelines on fishing seasons, catch limits, and protected areas, helps prevent overfishing and ensures resource sustainability. Engaging local communities through community-based management systems empowers stakeholders to actively participate in resource conservation and adopt sustainable fishing practices. Regular stock assessments and monitoring programs provide critical data for informed decision-making, preventing the depletion of fish stocks. Preserving and restoring aquatic habitats, such as wetlands and mangroves, is fundamental to maintaining a suitable environment for fish breeding and sustenance.

Adopting selective fishing gear minimizes bycatch and reduces the impact on non-target species, contributing to the overall health of the fishery. Integrating sustainable aquaculture practices, including polyculture and integrated farming systems, enhances productivity while minimizing environmental impact. Education and capacity-building initiatives ensure that fishermen and communities are well-informed and equipped to implement and adapt to sustainable measures. Recognizing and addressing the impacts of climate change on inland fisheries is crucial for building climate-resilient management strategies. Additionally, promoting sustainable practices through certification programs and eco-labeling incentivizes responsible fishing, creating market opportunities for sustainably sourced fish products and supporting economic viability. Incorporating these sustainable practices into inland fisheries management not only meets socio-economic needs but also contributes to the preservation of ecosystems and biodiversity. This holistic approach ensures the enduring viability of inland fisheries, supporting the livelihoods of communities and fostering the conservation of aquatic ecosystems.

Future Directions for Inland Fisheries in India

Looking into the future, the trajectory of inland fisheries in India is marked by optimistic prospects and a strategic vision for growth, sustainability, and socio-economic development. Several key directions and strategies are emerging to guide the sector towards a resilient and vibrant future. Embracing technological advancements is a cornerstone for the future, with a focus on integrating cutting-edge technologies in aquaculture practices, data analytics, and precision farming. These innovations are expected to enhance productivity, optimize resource utilization, and contribute to the overall efficiency of the sector.

Sustainable aquaculture practices will continue to be a driving force, with an emphasis on eco-friendly approaches such as recirculating aquaculture systems and integrated multi-trophic aquaculture. These practices aim to minimize environmental impacts and promote responsible resource management. Adapting to climate change is a priority, necessitating the development and implementation of climate-resilient strategies. Fisheries management approaches will need to evolve to address the uncertainties posed by changing environmental conditions, ensuring the sector's adaptability. Community participation and empowerment will remain a central theme, with a focus on involving local communities in decision-making processes, resource management, and the promotion of sustainable fishing practices. This approach strengthens the social fabric and encourages responsible stewardship.

Conservation efforts will intensify, with a focus on preserving biodiversity and restoring habitats. Programs dedicated to riverine and wetland conservation will contribute to the ecological sustainability of inland fisheries (Pandit *et al.*, 2019). Exploring opportunities for value addition and market diversification will be crucial. Promoting high-value and niche species, developing robust value chains, and connecting with global markets through certification initiatives are envisioned to enhance the economic viability of the sector. Establishing research and innovation hubs dedicated to inland fisheries will be instrumental in driving progress. Collaborative efforts between research institutions, industry stakeholders, and government agencies can lead to the development of innovative technologies and practices.

Continual policy support and effective implementation will be critical for realizing the sector's potential. Aligning policies with sustainable practices, providing financial incentives, and ensuring adaptive regulatory frameworks will create an enabling environment for growth. Investing in capacity building and skill development programs will empower the workforce, ensuring that fishermen and stakeholders are equipped with updated knowledge and skills in modern fishing techniques, aquaculture, and post-harvest practices.

Conclusion

The future of inland fisheries in India revolves around a holistic approach that integrates innovation, sustainability, community engagement, and resilience. By charting a course towards inclusive development and environmental stewardship, the inland fisheries sector is poised to play a vital role in India's sustainable and prosperous future.

Further Reading

Bennett, N. J., Schuhbauer, A., Skerritt, D. and Ebrahim, N. 2021.Socio-economic monitoring and evaluation in fisheries. Fisheries Research, Volume 239, https://doi.org/10.1016/j. fishres. 2021.105934.

Ekka, A., Katiha, P. K., Pandit, A., Barik, N., Shyam, S. S., & Ganesh Kumar, B. (2012). Socio-Economic status of fishers of reservoirs of India. Journal of the inland fishery Society of India, 44(2), 79-87.

Katiha, P. K., Pandit, A., Ekka, A., & Sharma, A. P. (2017). Socioeconomic status of riverine fisher communities in India. Aquatic Ecosystem Health & Management, 20(1-2), 188-197.

Pandit, A., Ekka, A., Das, B. K., Samanta, S., Chakraborty, L., & Raman, R. K. (2019). Fishers' livelihood diversification in Bhagirathi–Hooghly stretch of Ganga River in India. Current Science, 116(10), 1748-1752.

Index